High and Low

Keith Foskett

www.keithfoskett.com

High and Low

By Keith Foskett

www.keithfoskett.com

Copyright © 2018 Keith Foskett. All rights reserved.
ISBN: 978-1977926210

Cover image © Jeremy Rowley

Cover design by Mark Thomas, Coverness

The right of Keith Foskett to be identified as the author of this work has been asserted by him in accordance with the Copyright, Designs and Patents Act of 1988.

This book is sold subject to the conditions that it shall not, by way of trade or otherwise, be lent, resold, hired out, or otherwise circulated without the author's prior consent in any form of binding or cover other than that in which it is produced and without a similar condition including this condition being imposed on the subsequent purchaser.

This book is written using British spelling and grammar.

I love you, Mum

Depression

Depression is a state of low mood that affects a person's thoughts, behaviour and feelings. Common symptoms include lethargy, tiredness, irritability, loss of appetite and feelings of helplessness. It is a leading contributor to suicide.

Depression, in its various forms, affects some 350 million people worldwide.

Contents

Acknowledgements

It's you I want to thank first. Yes, YOU. Perhaps you're sitting in your lounge, maybe lying on the beach, or reading this on the train to work. Whomever and wherever you are, thanks. If it weren't for you supporting me, I wouldn't be writing.

I've had a habit in the past of trying to list people here in order of importance. Quite what this means or how I do it I don't know, because this is no place for a hierarchy.

You've all given up your time, and that makes you special (even the ones who charged me!). Once again, I couldn't have put this book out there without you.

In no particular order, my heartfelt thanks to:

Everyone who responded when I reached out for help with my depression, and those I didn't need to ask.

To the Street Team, the people who assist prior to launch with proofreading and associated favours. You're all fantastic!

Wendy Werneth, Cindy Todd, Ian Goulbourne, Ken Monaghan, Wendy Warner, Scott Norton, David Williams, Susan Morris, Susan Jackson, Nigel Higgs, Deborah Mutton, Bernard Duffy, Miss Hall, Jill Doyle, Norman Reeves, Hilary Brown, Marcus Loeffler, Barbara Ablitt, Tim Wolfers, George Thomas, Dave Fraser, Johnny Varner, Marcin Bajer, Amanda Haley, Charley Seger, Christine Saul, Craig Davenport, Liz Wannop, Stephanie Cooke, Frances

McCallum, Georgina Talbot, Richard Green, Steve Castle, Gregory Stephens, Nicola Ebert, Julie Haigh, Gary Ringrose, Polly Wise, Pip Fairweather, Glen Smith, Charlotte Collins, Michael Hewitt, Shaun Bades, Justin LaFrance, Paula Forman, Derek Fogg, Ayako Odashima, Jeremy Korn, Joseph Harold, Patrick Wadsworth, David Le Hunte, Megan Shumaker, Stephen R. Marriott, Nina Smirnoff, Kate Bryant, Simon Prendergast, Janek Mõttus, Justin Dyson, Troy Abbott, Izabela Walczak, Holly Ayers, and Mike Cunningham.

To Neil and Jannion at Routebuddy mapping software. I would have got seriously lost without you!

Chris Townsend for the foreword.

Alex Roddie at Pinnacle Editorial for work editing and formatting despite difficult circumstances.

Jeremy 'Obs da Blobs' Rowley for the cover photography.

Mark Thomas at Coverness for the cover design (nipple editing all part of the package).

James Thacker, Zac Poulton at MountainZ, Rich Cross at Alpine Guides, Chris McDonald at Scottish Canals, Ruth Hernández Paredes (President of the International Appalachian Trail Spain). Brad Devine – thanks for the email!

Introduction

Join Me on a Grand Adventure

Read this if you suffer with depression
If you don't, read it anyway

W hen I wrote this book, I incorrectly assumed it would be straightforward.

I thought the same about recovering from depression and undertaking this hike.

Wrong again.

If you're suffering with depression and expecting a miraculous cure, I'm afraid you've bought the wrong book.

However, there is hope.

I want you to join me on a grand adventure. This will entail 31 days of hiking across one of the most beautiful countries in the world.

We'll get lost, experience ferocious weather, the ubiquitous Scottish midge, strange-sounding local delicacies, and substandard

TV sets. Occasionally the sun appears. I'm a nice guy – come with me.

It will be fun, may surprise you, even shock you.

Only when you immerse yourself in this physical journey will you begin to understand the psychological voyage.

That mental discovery forms the second expedition, and contrary to what I said about being unable to offer a cure for your depression, I *can* help you. I can provide hope, and believe me, if you're hurting inside, *there is hope.*

Realise that, and you've already made the first step in restoring control of the torment that rages within. I know you've cried, hidden away, ruined friendships, and not only lost faith in your life, but also considered ending it.

Promise me this: if you are suffering from depression, don't turn to the back of this book. Witness my journey first, and when you do reach the end, you'll be in a better position to find out why I can help you live a relatively normal life.

Keith Foskett
Recovering depressive
Thursday, December 21st, 2017

Foreword by Chris Townsend

Long-distance hiking in Scotland isn't easy. The weather, the midges, the state of the paths (where they exist), and the rugged terrain all make walking challenging. People who've walked all over the world and done some of the longest trails come to Scotland and are surprised at how tough the hiking is. Imagine then how much harder it must be if you're suffering from depression. And on top of that, you've just had to abandon a major long-distance walk – the Continental Divide Trail in the USA – due to physical illness. That was the position Keith Foskett found himself in when he hiked the length of Scotland and which he describes vividly in this book.

How did Keith manage such a challenge? What kept him going through the days of rain, midges and mud as well as his illness? Having met Keith, and having read his books on his previous long-distance walks, I think it was his mental toughness and drive – something that is essential for completing such walks. Those past experiences, well described in his books, must have really helped. Even so, the walk must have been more difficult than I can imagine.

For those of us who enjoy long-distance walking, it becomes an addiction, a necessity, a major part of our lives. That's the case with Keith, and this desire, this need for the freedom of the trail and the power of nature and wild places, was clearly enough to carry him beyond his illness and keep him walking.

In *High and Low*, Keith describes his walk from Cape Wrath right down Scotland to Kirk Yetholm. That's not an easy walk, especially the first section to Fort William. I know this country well, and I know just how hard the walking can be, especially when it rains for days on end. Navigation isn't easy either – there are few signposts. Unlike Keith's other long walks, all on established popular trails, there weren't many other hikers going the same way to socialise with, to share hardships with, or to provide moral support. A walk like this requires the ability to be happy alone and to have the mental strength to deal with difficulties by yourself.

As well as the details of his walk, Keith captures the feeling of being on a long walk well – both the joys and the tribulations. His sense of humour comes through too, and he's not afraid of laughing at himself. This makes for an entertaining book that also has a serious undertone. It's a welcome addition to the literature of long-distance walking.

Chapter 1

Biological Breakdown

19th May
The Continental Divide Trail, New Mexico, USA.
Mile 446 of 3,000.

A heart attack at 10.30 in the morning isn't a situation I had planned for. I sat on a rock, wincing as the exertion increased the piercing pain near my heart, which thumped aggressively as if trying to escape. Out of instinct, I reached for my rolling tobacco.

That's not a great idea, Fozzie.

I paused and reconsidered. The day had started fine; waking at 6am, I'd boiled some water for a strong coffee and thrown some dry granola down my throat in between packing my gear up. I'd camped in a hollow, surrounded by towering pines. The sun struggled to make any inroads, peering around the trees shyly as it spliced through a weak mist. I was away quickly, eager to warm up as I sank my chin deeper into my jacket to gain a couple of degrees' advantage.

An hour later I ignored the minor chest pain, assuming that my body was just stretching and ironing out any weak spots as normal. But, gradually, the pain worsened. My breathing laboured, unable to pull in a full lungful of the dry, desert air, I stopped short of a full breath, and struggled to exhale. I knew I should stop, but I pressed on, my thoughts focused on Cuba, the next town which I needed to reach for a food resupply. Biological breakdown didn't feature on the master plan and would have to wait.

I shifted position on the rock and considered my predicament. Grants, the last town, was 36 miles behind me, at least a day and half's walk – but the problem was that I couldn't walk. Cuba was 70 miles ahead. I was aware that there were hikers coming down off Mt. Taylor, visible to the east, because I had spoken to them in Grants two days prior and knew their plans. The junction where my trail met theirs was miles ahead. I clutched my chest and rocked back and forth. Someone needed to come up the trail, and it needed to be quick. I haven't been worried many times, but this time I was scared. I feared for my life.

As if my plea had been answered, I heard footsteps and looked up.

"Hey! What's up?"

Hojo's face dropped when he saw mine, not bolstering my confidence.

"My chest hurts like a bastard, Hojo, and it's getting worse."

I had started the hike with Charlie 'Hojo' Mead some four weeks earlier. We had met five years ago on the Pacific Crest Trail and stayed in touch, perhaps knowing that someday we would both take on the CDT, and it would be ideal to hike it with someone we knew and gelled with. Hojo had left to take the side trail up Mt. Taylor a day before me, so I expected him to be way ahead. Of all the people who could have happened upon me, it couldn't have been any better. He worked as a ski patrol medic, probably the next best thing to a doctor.

"Where does it hurt?" he asked and dropped his pack quickly.

I pointed just by my heart, struggling to straighten myself and sit upright.

He held my wrist to take a pulse. "How's the pain on a scale of one to ten?"

"About a seven," I offered.

"Getting worse?"

"No, it was hovering around an eight but seems to have stabilised."

"Any pain, numbness, loss of sensation around the area? Your arm? Neck?"

"No. I can't take full breaths though, my breathing's restricted."

He checked his phone whilst meeting my gaze directly.

"I haven't got a signal," he said, looking slightly anxious, "Have you?"

I glanced at my phone. The signal indicator, as if undecided, flickered between nothing and one bar.

"Yeah, but it's weak, really weak. What do you think?" I sought reassurance, returning his gaze.

"You've got chest pain. You don't fuck around with chest pain, we need to call 911. Stay sitting, try and relax."

He took my phone, holding it aloft in some vain attempt to improve the reception, and walked down the trail.

"Hojo! Wait! Here." I handed him a scrap of paper on which I had scribbled our location coordinates.

Ten minutes later he returned.

"OK. Somehow I got through to the sheriff. The line was

poor but he said sit tight," he paused. "I wouldn't do that if I were you," he added, seeing me pull a cigarette paper. Scribbling more notes, he took my pulse again and asked about the pain score.

"It's going down, maybe a six. It still hurts to breathe."

We waited by the side of the trail for an hour, mistaking the occasional breeze rustling through the trees for a truck engine which I hoped would come bouncing down the rough track any minute. Then, an unexpected sound – *whoop, whoop, whoop.* Hojo stood up, shielding his eyes from the sun, and smiled.

"Looks like you got a ride in a chopper!"

The helicopter appeared from nowhere, racing towards us as, suddenly, the nose lifted to arrest its momentum. It hovered, then circled us twice as Hojo stood firm, making arm gestures. Eventually it landed in the scrub a couple of hundred feet away, and the door swung open as two medics jumped out, crouching under the rotors to come racing towards us. I felt a little excited by all the attention.

Velcro ripped as various pouches and bags were torn open. Hojo relayed his notes to one of the medics whilst the other attached enough wires onto my chest to power a small village. A tiny screen flickered to life, and she scanned the information.

"Your heart's strong, really strong," she said, looking at me. "I don't think you're having a cardiac arrest or anything but we need to get you out of here. You going to be OK in the chopper? You OK with flying?"

"Yes."

Before I knew it, I was sitting upright on a stretcher, looking out of the window as Hojo vanished, a vague blob in the vast, barren New Mexico desert. Faint tracks cut across the wasteland, speckled with occasional clumps of trees, confirming just how far from civilisation I was. The medic occasionally placed her hand on my shoulder with an expectant look on her face, searching for a response.

"I'm good," I muttered through the microphone.

Bouncing and swaying, fighting wind currents, the chopper's compass settled on an easterly bearing.

"That's the Rio Grande!" the pilot said, gesturing downwards. "One brown, murky river!"

Rising sharply, we crested a hill to be met by the imposing sight of a city suddenly engulfing everything.

"Where are we?" I cried.

"That's Albuquerque! We're going to drop you right outside the Presbyterian Hospital, they'll look after you."

I awoke expecting, as usual, a view of New Mexico from my tent. Instead, a pile of discharge papers lay under my right hand next to a bottle of antibiotics; an empty pizza box and the TV remote rested on my leg as the weatherman silently pointed to storm clouds on the TV. I rubbed my eyes, squinted at the light streaming in through the window, piecing together what had happened.

"You need to rest, Keith. Bronchitis isn't something to mess around with. Don't do any strenuous exercise for at least three weeks. Oh, and you need to stop smoking!"

The consultant's words echoed in my head, bouncing off my skull and whirling around, constantly repeating.

Don't do any strenuous exercise for at least three weeks.

I reached for my tobacco.

That's not a good idea, Fozzie. Take a shower, go get some breakfast.

Sometimes I wonder whether the voice in my head is

actually mine, or someone else looking after me, guiding me through life. Whoever it is, they always speak the truth.

The diagnosis didn't really worry me – I'd recovered from worse. My concerns centred on this voice, and its reaction after the consultant had advised I get off the trail:

Thank fuck for that.

Scraping together enough money to thru-hike for six months, planning the expedition, and leaving England to tackle the 3,000 miles of the Continental Divide Trail was a dream I had nurtured for three years, ever since finishing the Appalachian Trail. The Pacific Crest Trail was already a notch on my hiking belt, and this adventure was the last of the big three hikes in the States. A triple crowner, the status bestowed on any thru-hiker who has completed all three trails, after around 7,850 miles of hiking in some of the most remote and extreme terrain anywhere, was waiting for me. I had failed, or at least my body had.

I live for hiking, relish every single, sweet moment of it. So, when the situation dawned on me after hearing the doctor's news, *thank fuck for that* wasn't the first thought I had expected to pop into my head. Surely it should be disappointment, sadness, a resignation of failure, at least until the next attempt? In truth, however, I *was* relieved – I was glad that my hike was over. After coming to terms with the fact that I could have died from a heart attack or bronchitis, miles from anywhere, my next struggle centred on why I was glad to be off the trail.

My mother had commented before I left that I didn't seem right, and I knew I wasn't. My head was spinning, confused, torn between leaving for the hike and battling a disturbing awareness that, for reasons unknown, I shouldn't be leaving, that the Continental Divide Trail was a bad idea. Of course, I fought my confusion and doubts, trying to ignore them. Why the hell would such doubts have any bearing on my adventure? I was leaving to hike 3,000 miles on one of the most revered hiking trails in the world, so this should not be a reason to cause concern. The reason for my hesitancy was simple, although I never told anyone at the time: this inner voice was screaming at me.

Don't go! This is not the time to do this!

Every five years or so, my life catches me completely off guard and my inner voice starts screwing around. I have come to accept these phases, and, despite numerous initial doubts, I treat them seriously. Each phase centres on the suggestion of a major life change, regardless of whether I think I am happy at that precise time or not. These are not, initially, conscious decisions. My inner voice proposes that I'm unhappy and, to resolve this, suggests I make some alterations, presents some hare-brained idea from nowhere, lays the cards on the table and demands a decision. As ludicrous as the cards seem, and despite my protestations, they gradually start to nibble away at my sanity until my hesitation dwindles and, most of the time, I happily accept this new idea.

Some years ago, this voice proposed that painting and decorating would be a great occupation. I think being self-employed was actually the prime attraction, but decorating was the only skill that I really knew how to do well, and hence had the most chance of success at. I accepted the suggestion, worked hard, built up a decent customer base, committed to carrying out a great job for each client, and made sure the entire venture succeeded, which it did.

Then, a few years later, I became disheartened with

getting covered in dust and paint every day. But I continued to kid myself that this was what I should be doing, despite my voice popping up again and telling me to get the hell out. I battled it for months before eventually conceding that I did, actually, loathe my occupation. Even then it was out of my control for a while, because there wasn't much else I could do to earn the same income. I blamed my inner voice for getting me into that area of work to start with.

You suggested decorating five years ago! Now you want to change your mind again?

However, the voice of wisdom was correct, so when it suggested hiking and writing, I started to investigate. The allure was so attractive that investigations didn't take long, and I gladly dived headlong into the change. Whilst I continued to decorate, I gradually spent more and more time feeding my passion, in the hope that one day it would become a success and allow me to stop decorating completely.

So to suddenly be faced with another impending phase of my life, where my head was trying to convince me that my thru-hiking passion was just another short-lived journey, threw me into turmoil. I did love thru-hiking, and I loved writing about it. I lived for spending several months in the wild, then returning to write a book about the adventure. The overriding desire was to get out of decorating, constantly repeating each day doing something I hated. I wanted to earn a passive income from being an author. Basically, I wanted to be free, or at least as free as I could possibly hope to be.

Thru-hiking and writing are my passions; they take priority over everything else in my life. To suddenly have my beliefs challenged, to have to face up once more to some higher power steering me out of another phase of my life, wasn't something I was prepared to put up with again.

This time I decided to fight.

Resigned to my fate, and the doctor's advice, I checked out of the motel in Albuquerque and caught the Greyhound north to Boulder, Colorado, where the rest of my belongings I had brought were with my friend Tania. I stayed with her for a day, feeling very sorry for myself and sucking up her sympathy, before catching a flight back home to the UK.

For three weeks at least, my hiking career was on hold. I just hoped that, for once, the voice in my head was wrong.

What I had failed to realise, simply because I was completely blind to it at the time, was that, although this inner voice wasn't exactly wrong, it had not been entirely upfront. Perhaps it had the facts but didn't want to burden me with the truth. Perhaps it just wanted to protect me.

In retrospect I think it was just trying to steer me home. To familiar surroundings, amongst family and friends, somewhere safe to be able to deal with a period of my life that I had no idea I was about to enter. Little did I know that the following year I was going to implode.

My inner turmoil had nothing to do with any lack of passion for hiking, or indeed unhappiness with decorating. I was and had been suffering from depression for years.

And I'd been completely oblivious to it.

Chapter 2

Haggis, Clapshot and Crappit Heid

T*here's talk of werewolves on the trail!*
The opening line of my friend Alex's email caught me by surprise. *Werewolves* was not a word that immediately sprang to mind when I enquired how his hike on the Cape Wrath Trail (CWT) in Scotland was progressing, but it certainly lit the spark of curiosity.

It was June, two weeks after my return to England. I had rested, and then rested some more after further tests in the UK revealed I actually had pneumonia, not bronchitis. There were enough antibiotics floating around inside me to make any bacterial colony stop abruptly, scratch heads and seriously reconsider.

As I passed week two of resting, and going stir crazy unable to exercise, I started to research walks in the UK. Part of the reasoning was that, if my infection returned, then I didn't want to be marooned 10,000 feet up the side of a mountain in deepest New Mexico, with little chance of help. Also, I doubted my travel insurance premium with pneumonia as a recent addition would be favourable.

The other part was simply that I had never done any long hikes in the UK. My passion for distant lands, especially America, had grabbed me five years ago and never let go. I have no problem with this – seeing America at its wildest will never leave me, and I'm not done with it either.

This is true for many adventurers. Travelling to the far reaches of the globe, to areas we haven't visited before with weird languages, strange food and unfamiliar scenery, sparks the feeling of excitement we live for.

However, the downside is that few of us take the time to actually go and explore our own countries, thinking that, as we live there, we have all the time in the world; and except for short trips, we tend to leave them until last.

It's certainly true in my case. I always find it funny that I meet Americans and people of other nationalities in the UK who are keen to experience my country, whilst I'm on my way to see theirs. And when I'm hiking up a big mountain in the Sierra Nevada, the Picos de Europa, or wherever, I meet locals who convey their desire to visit my country. We all keep crossing each other until one day, eventually, the time arrives to set off to see our own homeland in depth.

Decision made, all I had to do was find a reasonably long hike in the UK, preferably as far from civilization as possible. Seeking the wild places pretty much decided my destination for me. The UK is not a big place – you can drive from the tip of Scotland to the south coast of England in a day. It's also getting crowded. Our population is high, and growing all the time. Essentially, to seek out our remote corners, you have to head to Scotland.

Sitting pretty at the north of the UK, Scotland has vast tracts of the outdoors that remain unpopulated and for several good reasons. Firstly, large areas are mountainous, and whilst they aren't huge in stature (they sit under the highest of them all, Ben Nevis at 4,414 feet), it can seem like the roof of the world when you're out there. The cities tend to congregate in the south, and although Scotland is dotted with towns and villages, the lie of the land means there are few of them. Much of it remains inhospitable.

Secondly, the weather isn't exactly friendly. I'll probably have the Scots in uproar at my temerity to even bring up the subject, which the Tourist Board will adamantly deny. However, it's definitely not beyond the realms of possibility that whenever you take a trip up north, the elements will conspire against you.

I've made around ten trips to Scotland, and it's rained on me for most of them. OK, so England isn't exactly a sub-tropical paradise, but we do fare slightly better. It's no secret that the further you travel from the equator, the more the weather deteriorates. Scotland sits on the same latitude as Norway, Sweden and Canada. Indeed, much of it is further north than Moscow. Think about that for a second – have you ever seen pictures of Moscow in the winter? They're not strolling around in speedos and bikinis up there you know.

Global location aside, the mountains also compound the issue. Much of the western side of the UK sits higher than the eastern side. Wales, the Lake District in England, and western Scotland are all home to the tallest hills and mountains. When fierce storms come barrelling in from the Atlantic, having crossed a few thousand unhindered miles, Ireland gets hit first. This somewhat tempers the storms that manage to carry onwards to Wales and the Lake District, but Scotland lies north of Ireland, and so the west coast gets smacked with the full force of these storms. On reaching landfall, they slam into the mountains, where they either stop or slow down, dumping most of their contents. Much of the western side of the UK is regarded as the wettest.

Thirdly, allow me to introduce you to the Scottish midge. At this stage of my planning, I'd decided I could happily deal with mountains. I generally curse them when I'm slogging uphill and breaking my knees on the descent, but they more than make up for it with the views and endorphin gratification. Despite all the evidence to the contrary, I also figured, foolishly, that my expedition was going to be timed to perfection and Scotland was about to enjoy their best summer ever. Seriously, it was going to be warmer than the Bahamas up there.

These minor inconveniences aside, there was no escaping the midges. From May to September, much of Scotland becomes infested with the Highland Midge, *Culicoides impunctatus.* They are found in upland and lowland areas, prefer wet, boggy or marshy ground and are most common in western Scotland – exactly where I was intending to hike.

If you're are familiar with my walks stateside, you'll know that biting insects generally love me. If I'm even in the remote vicinity of a colony of mosquitos, they'll wolf-whistle to all their mates, send out invites, and turn up a few minutes later with a six pack and selection of snacks.

"Hey! Party over at Fozzie's!"

"Nice! Good eating! Whereabouts?"

"The arms are great, apparently, but I'm going to try for some neck! Put the word out!"

I had no reason to believe that the midge was going to be any different. If there was one aspect in my favour, on all of my previous trips to Scotland, I never had a problem with them; they just didn't seem to appear.

I leant back in my chair, hands clasped behind my head, pleased with my research but not entirely convinced by the results. So far, the master plan entailed walking through a mountainous area, inclement weather and squadrons of midges. I could almost hear the droning.

All this and werewolves to boot. I needed reassurance, so I called Alex.

"I just finished!" he cried. "It was great! I'm in Durness for a couple of days, we may even cross paths if you're leaving soon."

"How was the weather?" I asked, my eyes narrowing for clues even though I couldn't see him. I cocked an ear for even the slightest vocal weakness.

"Yeah, OK. Had some rain but generally it was fine."

"Midges? How were the midges?"

"Saw a few, you get used to them."

"OK, I'm looking into the travel logistics, I'll call you tomorrow. Buy you a beer if we hook up."

I hung up and looked at my computer screen. Staring back at me was a photo of a broad ridge on the Cape Wrath Trail. It was stunning; sunlight streamed down through breaks in postcard-perfect white clouds, dappling the terrain in light and shade. Distant peaks illuminated, rivers twinkled down in the lowlands, and, above all, it looked deliciously wild.

Ignoring the negatives thrown up by my research, I set about balancing my concerns with some positives. To start, one of the wonderful things about Scotland is the amount of light it receives during the summer months. Towards the end of June, the sun is up just after 4am, and it doesn't set until around 10.30pm. Throw dawn and twilight into the mix, and you can still see what you're doing for a good hour outside these times. That's an enviable, UV-loaded seventeen and a half hours of light. Most Scots don't even experience night-time at that time of year; they go to bed when it's light, and get up when it's light as well. If ever there were a trade-off, a barter if you like, to counter the midges, then the sunlight card trumps everything. Of course, this was based on the dodgy assumption that there wouldn't be clouds obscuring any of it.

My destination was also looking financially frugal not only to get to, but to support me during my visit. Durness, the nearest village to the start of the Cape Wrath Trail on the north-west Scottish coast, is about as far north as you can go. I live a few miles from the south coast of England, and Durness was just shy of 700 miles away. Even with the hassle of public transport, coach fares had never been cheaper, so my goal of preserving funds, especially after the dip they had taken getting to America and back, was looking promising.

The cost of living is also cheaper over the border. Accommodation costs less, even during the summer months (I think they knock it down a little to offset the midge and weather problem), and, to cap off my financial planning, beer prices up there make breweries in the south-east of England cry.

Food was cheap as well, and, contrary to what most people think about Scottish cuisine, I think it's pretty decent. Plenty of solid, hearty root vegetables. A choice of seafood which is the envy of the whole world, in particular lobster and scallops. And haggis. Did I not mention that?

Haggis, in its traditional form, is sheep's pluck (heart, liver and lungs), minced with onion, oatmeal, suet, spices and stock. This delightful-sounding mixture is then encased in a sheep's stomach and boiled. Traditionally, it's served with *neeps and tatties* (mashed turnip and potato). I've had haggis on numerous occasions, and I like it.

However, it all goes downhill from there. The Scots have an alarming habit of dipping food in batter, and deep frying pretty much anything they think they can get away with. Chocolate is a firm favourite, in particular Mars bars, but other forms of confectionary to hit the oil include Bounty bars, Cadbury Creme Eggs and Snickers. If you don't have a sweet tooth, many outlets will happily offer a deep-fried pizza.

Other Scottish savoury delights don't do themselves any favours by their names alone. Crappit heid (seriously), a traditional dish popular with the working class due to its cheaper cuts, consists of a cod's head stuffed with oats, suet, onion and the fish's liver, which is then sewn shut and boiled. The liquid is often passed off as soup.

Finally, clapshot, sometimes known as clapshaw or Orcadian clapshot, despite sounding like a diagnosis from the sexually transmitted disease clinic, is actually a simple dish of mashed potatoes and swede. I'm happy trying most food, especially when I'm hungry after a day in the hills, so these Scottish delicacies didn't bother me. Strangely, I was actually looking forward to them.

Another advantage of my expedition was that I also had far less chance of bumping into anyone else. Anyone who knows me well enough is aware that I enjoy my solitude. I like a beer with my friends as well as the next person, but I have no issues whatsoever going for a week in the wilds without seeing another soul.

I twiddled around some statistics from the web. Scotland is around 30,917 square miles with a population of 5,295,000. England is larger, but with far more people; 50,345 square miles, with 53,000,000 residents. I'm no mathematical genius, but a quick calculation told me that Scotland has 171 people per square mile, while England has 1,053. So, I figured I had far less chance of running into someone who could annoy me over the border.

I had successfully balanced the trade-off between Scotland's weather, big hills and midges with a generous amount of daylight (assuming I could actually see it), preservation of my hiking funds, decent food and getting away from everyone. Things were definitely looking up but I still had one area that needed work.

The Cape Wrath Trail starts, strangely enough, at the north-western tip of Scotland – Cape Wrath. It worked its way south towards the finish point at Fort William. The distance was sketchy, because the trail had several variants, but I was looking at anything from 200 to 260 miles, probably around two weeks' walking.

Basically, it wasn't long enough. I had no work commitments for several weeks and still a good couple of months left of the summer. The solution was obvious; I needed to extend the walk, and it was then that the idea hit me.

For years I had played with the idea of walking *across* Scotland. There are numerous incentives – apart from just going for a long hike – for a little goal setting, and they don't come much better than walking across a country. It is an impressive addition to anyone's hiking CV.

Returning my gaze to the map, the next step of the trip became quickly apparent. After reaching Fort William on the CWT, I could get straight onto the West Highland Way (WHW), heading further south for 96 miles to the town of Milngavie. Of all the trails in the UK, it is considered an absolute classic and is on the must-do list of most budding hikers. I didn't even question the option and chalked in the next stage of my adventure.

I'd seen portions of the West Highland Way on numerous occasions during drives up to Fort William, the town I'd used as a base for past hikes. Many times, when a mate was doing the driving, I'd peer out the passenger window, nose pressed against the glass, ignoring the rain trickling down, observing the trail a mile or so distant. I knew where it intersected the roads, and occasionally I'd catch a glimpse of a WHW hiker, wrapped up in waterproofs, covered in mud and usually scratching any exposed areas of skin that the midges had infiltrated. The route tinkers with the clear waters of Scotland's rivers and streams, wiggles through and around numerous lochs, dives in and out of forests, and bumps up and over several mountains and passes. It visits such iconic landmarks as Glen Coe, the Trossachs National Park and Loch Lomond.

Rumour has it that the trail is also well maintained. Being one of the most popular routes, if not the most popular route in the UK, it has far more foot traffic than most trails, and the Scottish authorities, no doubt keen to cash in on the influx of thousands of foreign tourists each year, want to keep them happy. The CW,T, on the other hand, is renowned for *not* having any form of trail at all. It entails sticking to a rough route, utilising whatever appears to be the line of least resistance. After 260 miles of that, I figured it would be a nice idea to get onto a national trail with a decent surface and regular pubs.

Now, at this point I was still on the western side of the country and had to turn east when I hit Milngavie to begin the actual cross-country part of my plan. Both the CWT and the WHW are oriented north-south. By the time I was due to arrive in Milngavie after completing both trails, I wouldn't have actually made any inroads into walking *across* Scotland at all. It was, admittedly, a major flaw in my planning, but quickly resolved.

The UK is littered with footpaths and bridleways. Check any map and there are miles upon miles of dotted red lines criss-crossing the entire country. I know from experimenting that it's possible to pick two points on a map and, with a little thought, walk from one to the other utilising just these rights of way.

I scanned the map once more and diverted my gaze over to the east side of Scotland, and everything in between. Sure enough, paths abounded, including, to my delight, the chance to follow long sections of the Scottish canal system, in particular the Union and Forth & Clyde canals. This opportunity scored major bonus points for my expedition. First, and most obvious, canal towpaths are flat. I feel no need to explain this any further.

Second, the canals have been there a while – nearly 200 years in fact. The little businesses that sprang up to service the industry and all the old buildings, bridges and locks are a crucial part of Scotland's heritage, and they were sure to bring a smile to my face as I followed the journey along the water. A walk through the past if you like.

Water itself is another bonus. Canal and river walks are popular, and for good reason. There's something wonderfully relaxing about a stroll next to a river, canal, or the sea. It's been proven that we find the sound of water, and even just being in its vicinity, relaxing. I indulge myself back home in West Sussex often, where the Wey South Path follows the old canal system from Guildford to Amberley. It runs less than a mile from my house and has provided a thoughtful escape many times.

I could follow the Scottish canal system virtually all the way across, where at some point I could veer onto a southerly bearing and finish at Kirk Yetholm, not far from the coast and a little way from the English border.

I called Alex.

"When you leaving Durness?" I asked.

"Thursday morning."

"What's the weather doing?"

"It's great! Sunshine."

"OK. I'll be arriving Wednesday afternoon. Where you staying?"

"The campsite next to the Sango Sands Oasis bar."

"I'll meet you in the bar at, say, seven?"

"See you there!"

I made some coffee and, with droopy, tired eyes, summarised my plans. As a pastry chef might carefully measure out their ingredients into a mixing bowl, then stir, bake and rest a cake, I did the same with my ideas. I was hoping for moist carrot cake, with plump raisins and a few spices, thoughtfully finished off with one of those cute little iced carrots on top.

To start, I had a 260-mile route with no trail whatsoever, situated in the direct path of some pretty nasty Atlantic storm fronts. Numerous mountains and passes needed to be conquered, abating somewhat when I hit the West Highland Way, where I'd be crowded out by hordes of camera-wielding tourists. My course then turned east, taking advantage of the prevailing winds and, most likely, rain as well. All the time, I'd be subsisting on boiled sheep's stomach stuffed with its innards, and mashed vegetables. Doubtless, I'd be covered in rashes and scratching desperately whilst fleeing from clouds of midges. If I managed to get through all of that, the whole expedition could be brought to a bloodthirsty end at any time by a rampaging werewolf.

It was looking more like a serving of crappit heid than a slice of carrot cake.

But I liked it.

Chapter 3

The Humble Bothy

"**W**elcome to BT answer 1571. The person you are calling is not available. Please leave a message after the tone."

"Hi. My name's Fozzie and I'm walking the Cape Wrath Trail tomorrow, beginning from the lighthouse. I need to know if the firing range is in use then, because ideally I'd like to start with all limbs intact, and preferably alive. Would appreciate someone returning my call please. Thanks."

I hung up.

As far as a first day on trail went, this was new to me. The area around Cape Wrath is used as a naval gunnery range, and I'm not talking popguns either. Ships and planes let rip with artillery shells – the sort of ammunition that doesn't just pierce a leg, leaving you hopping to hospital. Those bad boys will pretty much incinerate anyone before they even know it's been launched.

Alex mentioned this rather important fact prior to my departure, and I had jotted the telephone number down.

Making the call en route to Scotland on Wednesday, I was somewhat miffed to be met with an answerphone. This wasn't my local bus station, and I wasn't trying to find out departure times; I needed to know my chances of getting vaporised. Hardly the time to leave the office unattended with the answerphone switched on.

I stared out of the window, lost in thought, my gaze unfocused. The 'luxury' coach bumped along unceremoniously, making alarming scraping noises every time we hit a pothole. For once, my relationship with public transport was running smoothly. I'd caught a train to London, then endured a sixteen-hour coach trip to Inverness, where I'd holed up in the youth hostel overnight. My luxury bus had departed to Durness the following morning. Public transport invariably screws up when it knows I'm coming. Either delays suddenly spring from nowhere, or logistics break down should I manage to actually board. If you ever see me waiting for a train or queuing for a bus, I'd recommend you go buy a coffee, read the paper, and catch the next one.

There seems to be a lack of north-south roads in Scotland. Any trip on these bearings entails using roads that are east-west oriented, with a few exceptions. Couple this roundabout route with the need for the bus driver to stop at most towns and villages to have a thirty-minute chat with the local fisherman, then deliver the mail box and a crate of lobster, and a fairly straightforward journey ends up taking a lot longer than expected. However, it was worth it just to take in the scenery and wind down a little, and I was enjoying the relaxed way of life typical of west Scotland.

There's a fierce (although I like to think friendly) rivalry between the English and Scottish, and numerous jokes as well. We tend to rib each other at every opportunity, and I'm no different. However, make no mistake, I have to concede that Scotland is absolutely drop-dead stunning. With a mesmeric mix of mountains, lochs, rivers, forests, and pristine beaches, all bathed in a soft northern light, it never fails to inspire me.

The coach negotiated one last corner, and, as the road dipped, Durness finally came into view. Just a cluster of buildings about 13 miles east of Cape Wrath, it is the nearest settlement to the start of the trail itself. Set sail north from there, and you won't reach landfall until the Arctic. Durness sits right on this north coast, where a few houses and some amenities cater for the locals and visitors. A few restaurants, a pub, a couple of local stores, a post office and a campsite sum up the place.

Areas of grass bordered by low stone walls gently shelved down to the North Sea where, with care, a path or two took me down to Sango Bay. Triangular rocks protruded through the sand like sharks' fins as the beach melted into turquoise waters, themselves gently blending to teal skies. Friendly clouds raced across the sky, and teasing light games made the North Sea glisten, the contrast of light and shadows constantly changing.

The sign on an old, weather-beaten gate directed me left with the aid of a helpful arrow.

'Welcome to the bunkhouse, please book at Mackay's Hotel next door.'

Being the middle of the afternoon, the hotel was quiet. I entered and waited at the counter, pondering whether to ring the bell before a lady appeared and checked me in.

"Hi, I'm Keith. Just need a bed in the bunkhouse for tonight please."

"Hi Keith, I'm Fiona and of course, no problem. It's all straightforward," she said. "This key is for the front door and your room is the second on the right. It's quiet – you may be the only guest staying there tonight."

"Thanks."

I swung the gate open, climbed the stone steps, and thudded over the wooden deck, then opened the front door and found my room. I get vertigo in top bunkbeds and fell out once in the early hours trying to go for a pee, so I made claim to the bottom bed by scattering a few belongings over it. Retrieving my phone and wallet from my backpack, I left to explore Durness.

It appeared sleepy. An occasional campervan rolled past, the passengers peering and pointing out of the windows. Locals strolled with their dogs. A few walkers, standing out in lurid waterproof jackets, ambled towards the campsite, where an equal mix of tents and caravans speckled the grass. A car waited by the petrol pump, and the tourist information office was visible down the road.

A sign directed tourists to Smoo Cave, a complex of caves accessible either by boat or by a path from the cliff above. Evidence of human occupation dating back 5,000 years has been found in the area. The Picts – tribal people who lived in north Scotland around 2,000 years ago – also left their mark, and archaeological excavations have discovered remains of their farms.

I walked through the campsite and found a small knoll, where I sat and stared out across the ocean, contemplating the next few weeks. I get a lot of time to think on any hike, and no moment proves more contemplative than the start. I wasn't due to begin until tomorrow, but already I was at peace, rolling around my next adventure. I was convinced my journey was going to be a wild one. Although never really that far from civilisation, Scotland should never be taken lightly – it can be a dangerous place. In between the resupply points lies the wildest corner of the UK. I never underestimate the outdoors; it doesn't matter where I am in the world, from the middle of the Californian desert to southern England. I've learnt through experience that complacency is an enemy not to be underestimated. I've been caught out in New Mexico, and, even on my local hills, the South Downs, I've been faced with challenging experiences.

However, danger and the unknown that our wild places bring, as any adventurer will confirm, are why we go there in the first place. New Mexico returned to my thoughts, and I realised how lucky I'd been. I had truly believed that I was dying and that someone was going to happen upon me a day later. It is these dangers and fears that drive many of us – we feel alive out there, and situations where we are truly scared teach us how rewarding it is to be experiencing and appreciating not only the adventure, but life itself.

I could just make out Alex's silhouette as I entered the Sango Sands Oasis, backlit by light streaming in through the window that bounced off a polished mahogany table and ricocheted straight through a very inviting pint of amber bitter.

Alex and I have much in common. He is also mad on hiking and the outdoors and is also an author and blogger.

"How are you, chap?" I asked, shaking his hand.

"Great!"

He was beaming, no doubt pleased at having just completed what is regarded as the toughest hike in the UK.

"How was it? Fill me in."

Alex recounted his adventure animatedly. Two thrill-seeking hikers getting more excited by the minute, I don't know who was more enthusiastic, Alex having finished or me about to start.

I listened as he relayed the past couple of weeks and began to realise what I had let myself in for. The CWT was no pushover and was about as difficult as a hike could get. Alex had managed to persist through several days of appalling weather, battling a trail which, at best, consisted of bog. Navigation had been difficult, and he warned me that, at day's end, finding anywhere to pitch a tent was a battle due to the uneven, wet ground. Points to resupply with food needed to be planned for. I didn't feel like broaching the midge topic.

I had to stop him at one stage, having visions of struggling up the side of a mountain, battling a howling gale, with no food, and water up to my ankles. There was only one thing for it: I ordered another pint and a whisky chaser.

Alex retired to his tent, and I wobbled back to the bunkhouse, realising that he hadn't entirely been up front with me on the phone.

Digesting a hearty cooked breakfast in the morning, I called the Ministry of Defence again, but hung up quickly when the answer machine kicked in once more. I decided to leave anyway, figuring I could at least stay somewhere near the lighthouse that evening until I knew it was safe to walk. There was a café there for the few visitors who ventured up to Cape Wrath, and I hoped they might have news on any impending bombardment. Then my phone rang.

"Hi Fozzie, we got your message. The firing was due to continue into the weekend, but it's been called off. You're safe, have a nice walk."

I smiled, shouldered my pack, adjusted my laces, and started to walk to the Kyle of Durness, an estuary about an hour down the road. It was a bright day again, and I pondered if my luck with the weather would hold out. The single-track road was quiet, and I walked slowly in the middle, squeezing onto the grass verge for the occasional car. Turning right after a couple of miles, I took a narrow side road down to a quay where seven people were boarding a tiny boat, filling it to capacity. I unclipped my pack, laid it down, and sat, waiting for the ferry to return about an hour later. A solitary anchored boat floated motionless on the Kyle, a wide, sand-bordered estuary fed by the Dionard and Grudie Rivers. Rugged peaks rose above the inlet, a stark contrast to the smooth banks of sand shelving into the water. A red buoy offered the only splash of colour as clouds, certainly far darker than when I had left, reflected perfectly in the mirrored surface. I watched the ferry carve a route to the other side, forming a white, v-shaped wake behind it, and could just make out the road winding up and disappearing over the distant hills. A blue minibus waited by the jetty, ready to take the passengers up to the lighthouse at Cape Wrath.

I waited, at times impatiently. I guess I still had a little winding down to do, but eventually the minibus returned, bumping unceremoniously down to the quay. I boarded the boat, and fighting a rising tide we finally reached the other shore, lurching forward as the throttle eased. I disembarked and climbed aboard the minibus, which is imprisoned on this lonely stretch of potholed road forever.

I became restless with the constant chatter of other passengers, and my stomach gurgled at the thought of a cheese and pickle sandwich at the Ozone Café, just by the lighthouse. We pulled up about an hour later.

Crappit heid didn't feature on the menu, and, despite my interest in sampling the local delicacy, I decided that getting through Scotland without trying it might be a blessing. The bus driver leant on the bar in the Ozone, framed by a small hatch.

"Is there anyone about?" I asked, becoming hungrier by the second.

"He's probably having a nap," he replied. "He tends to nod off when we arrive."

Ten minutes later the proprietor appeared, rubbing his eyes and offering a weak smile.

"Hi. Er, a cup of tea and a cheese and pickle sandwich please."

I wolfed it down, shook the hand of an alarmed-looking tourist who had enquired where I was walking to, then made my way to the lighthouse and the start of my adventure. I could just hear the muffled sound of water breaking against the base of the cliffs below. The hills rose and fell away to the distance, and I strained my eyes to pick out Sandwood Bay, my first destination, but the hills obscured it.

Right then Fozzie, here we go again.

It might seem an oversight that I didn't have a paper map, but instead I had opted to download the route to my smartphone. I hate maps, and the purists will scorn me for this, but I got fed up a long time ago with the folding issues, amongst other annoyances. If you've ever tried to fold an Ordnance Survey map then you have my sympathy, and I hope I have yours. If you're an origami master then rock and roll, but most of us aren't. Even on a clear, sun-bathed day with no wind, the innocuous OS map can reduce the best of us to tears. No matter how I study the folds and concentrate on carefully bending them the right way, map retaliation is inevitable. Try it in a gale at your peril. I had all of Scotland stored on my phone, which was conveniently located in a shoulder pouch, available immediately, and it showed me exactly where I was with the aid of a helpful little cross-hair icon. I guarded it with my life, ensuring it was safely stored with the pouch fastener firmly closed. I also had a battery storage pack for recharging, capable, I was assured by the manufacturer, of juicing up my phone a whopping seven times. I even had a spare phone, wrapped in a waterproof bag in a pack. Scared of getting lost? Me?

I swiped the screen and studied my position. My little cross-hair flickered, jumped about somewhat as if undecided, then eventually settled on my exact location.

This can't be right.

A mere 30 minutes into my expedition, and I appeared to be in the shit already. But the map, as always, was correct. A series of distinct red dashes signified the mile-long road I had just walked down from the lighthouse. Thanks to my pre-trip research, I knew that the Cape Wrath Trail left this road after a mile or so and veered off right, and south into the Scottish wilderness.

During this research, I also discovered that the map key depicting the route changed according to the condition of the trail. My path for the next couple of weeks alternated at a whim from, at best, smooth gravel to, at worst, bog. A series of firm, red dashes signified a great trail. As the path deteriorated, those red dashes changed to dots. Then the really rough stuff kicked in. A change from red to black was cause for concern. Black dashes not so bad, but black dots were reason to worry. And then, *then*, there were the occasions when the map didn't show any dashes or dots whatsoever, because there was no trail. I pondered why Scotland was promoting this route, and indeed why the map had it marked, when in fact at times there wasn't a trail at all.

That map was to play havoc with my emotions until Fort William, some 250 miles distant. I tried to spend the night somewhere where my map offered red dashes. This ensured I could kick off the day hiking on a good surface, and therefore put myself in a good mood. Once those black dots appeared, my mood slumped accordingly.

Imagine this: if someone were able to see my live position on their PC screen back home, they could very accurately gauge my state of mind at any given time. If my little cross-hair slowly inched along red dashes, it was probably safe to call me and enquire how things were progressing. Something like this:

"Hey Fozzie! Heard you were doing the Cape Wrath! How's it going up there?"

"It's amazing! Fantastic part of the world, great trail, amazing scenery! It doesn't get much better!"

Woe betide anyone who made that call as I hovered on black dots, or worse, no dots at all.

"Hey Fozzie, heard…"

"What the hell do you want?"

I checked the screen once more, hoping for a navigational malfunction, then focused on the terrain. A landscape of tussocks and surface water twinkled in the midday sun. Occasionally I could just make out lighter, brown sections of firm trail. I tightened my pack and left the road, descending in the direction of the west coast. Within minutes my feet were soaked. The higher tussocks offered islands of dry haven but were impossible to walk on, so I resigned myself to taking the path of least resistance. I can best describe the ground as one huge, several-thousand-square-mile sponge. Placing each foot on the ground, it slowly sank as the peat-tinted water rose and engulfed my foot with an unhealthy squelch. On occasion the sponge developed jaws as well, clamping a firm bite around my shoe as I lifted a leg only to see my footwear prised away to sink alarmingly into the sponge abyss, whilst I frantically attempted a rescue operation before it disappeared forever.

There is no trail!

Alex's words echoed in my head.

Some days this would go on for hours, the battle of the sponge only relenting as Scottish bog eventually hit terra firma and my world changed from a series of black dots to the calm oasis of solid red dashes. Never had a hiking surface affected my state of mind so much. Occasionally, the black dots nearly reduced me to tears whilst the red dashes made everything great again. I even named them. Black dots were known simply as 'dystopia', because it wasn't a place where I enjoyed hanging out. Red dashes were christened 'the path of limitless Guinness' because my mood whilst following them bore some resemblance to watching, in anticipation, a bartender skilfully pouring a perfect pint of the black stuff.

After successfully negotiating my first stretch in dystopia, I reached a small stone shelter at the Bay of Keisgaig. On one side the west coast plunged down to the beautiful waters below as waves lapped on the shore. To the other rose the modest summit of Cnoc an Daimh. I placed my pack against the shelter wall, pulled on my wool hat as I chilled, and munched on some almonds. It was to be a short day. My destination, Sandwood Bay, had been on my must-see list for years and offered my first overnight camp.

I had a clear line up and over Cnoc a' Gheodha Ruaidh before descending to Sandwood. A path of limitless

Guinness proved easy going and, sure enough, despite a brief flirtation with dystopia, in late afternoon my world descended gently and spilled onto smooth, clean sand. I was alone, and a lack of footprints suggested I was the only person to walk on Sandwood that day, making it even more special than my imagination had hoped for.

The wet sand held firm. I strolled along, passing Sandwood Loch to my left as this pristine beach stretched away for a mile ahead of me. On clear days with plenty of sunshine, you could mistake this little corner of the UK for somewhere in the Mediterranean. The sand seems almost white (some even claim it takes on a pink tint), and the waters appear turquoise, befitting any Greek beach. Or perhaps that should be befitting any Scottish beach? Visitors to this corner of Scotland travel miles just to see Sandwood Bay and to take a walk along it. Am Buachaille, a sea stack of Torridonian sandstone, pierces up through the sea at the southern end, framing the entire spectacle perfectly.

I removed my shoes and socks. The sand gave way slightly with each step as it squeezed between my toes. I glanced back to see footprints follow me, each gradually vanishing, succumbing to molten sand.

This beach stays quiet, and, if you are lucky enough to visit, there's a reasonable chance you will have it to yourself. The closest place to park is Blairmore, four miles away, and the approach track will take most tourists up to two hours through peat moor, speckled with wild flowers in the spring and summer. It's not a huge walk by any stretch, but most won't take it on, leaving Sandwood gloriously quiet.

It also has a colourful history. There is evidence of an early Pictish settlement, and it is believed that the Vikings landed here, dragging their boats over the beach into the loch. In fact, the name Sandwood itself is thought to be derived from the Viking name 'Sandvatn', meaning 'sand water'.

On the 30th of September 1941, Sergeant Michael Kilburn from 124 Squadron crash-landed his Spitfire on the beach after engine failure. The course of time, and seawater, has eroded the aircraft, but at low tide the engine is still visible.

Until the Cape Wrath lighthouse was built in 1828, many a ship came to an unceremonious end here. As I paused to gaze across this gracefully arching beach, it was hard to imagine how somewhere so tranquil and innocent had claimed so many lives. Shipwrecks are still thought to lurk under the sand, and locals tell stories of buried Viking longboats.

The bay is renowned for its mysterious past. Legends, ghost stories and perplexing stories abound, and reports of visitors feeling uneasy are common. Now I love a good mystery. If there is even the remotest whiff of the unexplainable, ghost sightings, UFO abductions or even werewolves, then my ears prick up and I sit expectantly, waiting to be filled in with the details like a puppy waiting for food. So, apart from a desire to visit Sandwood that stretches back many years, purely to lay eyes on it, I was hoping to experience something out of the ordinary. Perhaps some marauding Vikings sharing my supper or a strange apparition floating across the dunes late evening. I was curious no doubt, but it was giving me the willies just thinking about it.

Sandwood House, once a bothy but now a ruin, is the focus of many stories. Plenty of visitors have reported loud banging, footsteps, tapping on the windows, the sound of galloping horses and even the ground shaking. One particular ghost, a bearded man, thought to be a sailor, has reportedly been seen many times. He seems to roam at will, sometimes spotted just standing on the dunes staring out to sea, at other times walking on the beach or frequenting the area around the bothy. Sandwood Bay is a place of conflicts; dark legends, mystery, and death gently tempered by a mesmeric beauty.

If you're not familiar with bothies, let me fill you in. They are very basic shelters, and, whilst there are a few in Wales and England, Scotland has the most, around a hundred.

Bothies fall well short of five-star accommodation, thankfully. With very few exceptions there is no running water or electricity. At best you will have one, possibly two rooms, a fireplace if you're lucky, and a raised wooden platform to sleep on. There might be a table and chair. They are usually left unlocked, as the main purpose is to provide shelter from inclement weather, usually overnight.

I debated bedding down in the bothy, but nerves got the better of me. Carrying on to the southern end of the bay, I walked into the dunes and found a flat spot, then began to quickly erect my tent as light rain speckled my forearms. From the light cloud and sporadic sunshine earlier in the day, an oppressive sky had changed Scotland's mood, and now she sulked, grey and depressed. I dived into the tent, closed one of the two storm flaps to shield a light drizzle, and cooked my evening meal in the vestibule. During rain respites I stretched my legs as a few midges droned, and, finally, around 10pm, I could just make out a sinking sun peering shyly under the cloud base, finally crashing into the sea.

I woke at 6am with a bandana wrapped around my head. It was light at 4am, and, in a desperate attempt to regain some sense of darkness, I vaguely remembered tying it there to cover my eyes. It was a strange sensation going to sleep still bathed in light, then waking in light as well. This became a familiar pattern that repeated itself many times during my hike, and I began to doubt whether darkness in Scotland ever fell at all.

The rain had fallen sporadically all night, and, as I left Sandwood, promising I'd visit again, the skies hinted at more of the same. I passed many lochs, including Clais nan Coinneal and Meadhonach. After clipping the eastern edge of Loch a' Mhuilinn, I stopped counting them.

Pausing to climb over a stile at the edge of the firing range, curiosity got the better of me, and I peered at a sign by the fence, reading the faded writing.

> *The Ministry of Defence allows for public enjoyment of its estate wherever this is compatible with operational training needs, public safety and security.*
> *People must act responsibly as outlined in the Scottish Outdoor Access Code. The three key principles are: take responsibility for your own actions, respect the interests of other people and care for the environment.*

After reading it a couple of times just to make sure I hadn't misinterpreted anything, I started giggling.

In addition to the mainland around Cape Wrath, Garvie Island, which lies four miles east of the Cape, has been the focus of aerial bombardment for decades. Not just by the Royal Navy either; the Americans enjoy the odd pot-shot, as do some European countries. Indeed, it is the only range in Europe where land, sea and air training activities can be conducted simultaneously using live, 1,000lb bombs.

The net result of a lot of military personnel flying around slinging bombs willy-nilly at Cape Wrath, and Garvie

Island, is that much of it, along with the wildlife, has been decimated. There are craters everywhere and warnings of unexploded bombs, presumably because the army can't be bothered to come in and clean them up.

I looked at the sign again.

Take responsibility for your own actions. Respect the interests of other people. Care for the environment.

As hard as I tried to make sense of the hypocritical babble that some bored, pen-pushing general at the MOD had vomited from the bowels of absurdity, I couldn't. If anyone from the armed forces can explain how they are taking responsibility for their own actions, or respecting others, or indeed caring for the environment, please feel free to get in touch. I'd love to hear from you.

Despite occasional rain, I was happy. A path of limitless Guinness seemed set for the day, and at last I was back in the wilds, doing what I loved. My inner voice seemed silent, and on the odd occasions when it offered anything, the ideas were positive and focused. Despite previously suggesting my hiking days were over, I appeared to have won a stay of execution.

The morning brought overcast skies, but the cloud base was high, smooth and constant, tempered with swirls, ripples and eddies as the light glanced over. I'd never noticed this before; whether it was unique to Scotland I don't know, but it was mesmerising to watch. As the cloud drifted overhead, the density constantly adjusting, sunlight above played and danced. I sat many times just observing, wondering whether I was dreaming or this northern light show was for real.

Diffused light played across the landscape, shadows raced across the moors and charged up the mountains – it was constantly changing. In the evenings colours emerged with a richness that brought the bronze of the heather, the greens of the grass, and the grey of the rock alive. Stay awake long enough in Scotland, and your camera will love you.

I spilled out onto tarmac and the small cluster of houses that made up Blairmore, then picked up a narrow road that would take me a few miles to Kinlochbervie, rumoured to have a small store lurking in the industrial estate, of all places. I had plenty of food for a few days but never passed a shop by. First, it saved my supplies for the actual trail, and second, fresh food was always a draw. I was developing a dangerous addiction to either cheese or ham and pickle sandwiches, neither of which I usually ate – but long-distance hiking, even after one day, makes me develop all sorts of strange food cravings.

I turned right at a junction, the road dipping and skirting industrial buildings, a church appearing out of place. A local pointed me to the store, which was hidden up a small dead-end street. Everything was quiet; even the industrial area stood silent, and nothing moved. A solitary car was parked outside the store, which appeared shut until the door swung open and an elderly lady walked out as I approached. I paid for my sandwich and a bag of crisps, then returned to the church, where I leant against its cold stone wall and removed my sodden shoes and socks to allow my feet to breathe.

A narrow strip of sunlight crept through the picket fence and warmed my pasty, wrinkled toes. Every few minutes, as the sun moved and the fence shadow crept sideways, I adjusted my toes to keep in the light.

For most of the afternoon my feet stayed dry as I left Kinlochbervie and continued on the road for a couple of hours to Rhiconich, where I turned off just before an old arching bridge over the River Rhiconich.

The waters kept me company as the trail hugged the northern bank before funnelling out to meet the wider expanse of Loch a' Garbh-bhaid Beag. The converging waters narrowed before repeating the pattern, flowing into the bigger Loch a' Garbh-bhaid Mor.

I stopped to check what the map on my phone had to offer. The trail condition remained relatively good and followed a line of least resistance, heading south-west to squeeze in between contours and skirt around the surrounding hills. A small building called Lochstack Lodge was marked, which I decided to head for to bring in a 20-mile day. I didn't know if the lodge actually offered accommodation, or whether it was one of many fishing or hunting shelters, but it made a good target point to figure out my moves for the evening.

Despite a trail of limitless Guinness, my feet once more ended up soaked. Even on day two, despite not necessarily enjoying the feeling of constant wet feet, I had quickly accepted it.

Past Lochstack Lodge I came to a small wooden building. It was locked but appeared to be used by fishing parties, possibly guests from the lodge itself. It stood a few feet from the banks of the River Laxford and very conveniently provided a comfortable seat under the porch. I removed my pack, sat down, and decided to take advantage of the seating to cook my evening meal in comfort. Occasional drizzle created patterns on the river, flanked by acres of heather, as the imposing bulk of Meall Aonghais soared skywards behind me.

I had noticed a pattern emerging. The midges had been flying about since I started and, true to my research, tended to congregate by damp ground, especially the lochs and rivers. One trait the Scottish midge shares with most flying insects is that they don't like the rain because they can't fly in it. So, whenever I paused for a break, or stopped for the day, I had one of two annoyances to deal with. If it was raining I had to erect my tent quickly and dive inside for an evening of incarceration. If it wasn't raining, then obviously I couldn't get wet, but I would get bombarded with midges. It was a frustrating situation; get wet or get bitten. Either way, I spent a lot of time confined to my tent.

I started to learn, however, that although I couldn't control the elements, I could make life easier by picking camping spots that worked in my favour. I tried to stay away from water in the evenings and to select windy spots, as midges had difficulty flying in those conditions – a sure-fire way of keeping them at bay.

Despite this, sitting by the river waiting for some chicken noodles to do their stuff, I alternated between sheltering under the porch when it rained and running around in circles slapping myself when it stopped and the midges returned. Although not as big as mosquitos, they make up for their size in sheer numbers. I constantly swiped my face as they became caught in my beard. My ears, for some reason, were a favourite hangout (perhaps they considered them as I did porch shelters), and I had to slide my finger around my antihelix, collecting a reasonable colony of them. They also loved ankles, although I achieved limited success by tucking my waterproof trousers into my socks. I should point out that grey waterproofs, red socks and flip-flops (used as camp shoes to get out of my wet trainers) didn't do me many favours in the fashion stakes. Couple that with a silver poncho and a silver umbrella during rainstorms, and I looked like some sort of failed clothing trial bordering on space mission experiment.

Unable to find a dry tent spot, when the midges eventually got the better of me I retreated from the river, wounded in battle, to an abandoned building which appeared from the inside to have been used for keeping livestock. It was clean enough, and the roof didn't leak, but I had to stuff old newspapers into various holes to stop the midges infiltrating. One particular crack in the wall was too high to reach, and, although the midges hadn't discovered it, a new problem presented itself. As the wind whipped around outside and through the gap, my shelter echoed with the sound of moaning. To be exact, it sounded like the ghostly wail of a lost mariner, wandering around an old building late one evening, trying to scare the crap out of hikers. Every time the wind blew, a deep, soulless whine straight out of a Hollywood horror reverberated around the roof space. I pulled my sleeping bag over my head to escape the clutches of the lost seaman and prayed.

I was away quickly in the morning, aware of the need for an early start as the trail ascended. I crossed the river and skirted the flanks of Ben Stack to the west, thankful, at that time of morning, that the trail went around instead of climbing the mountain. I didn't escape Ben Dreavie, but, turning on an easterly bearing, I made short work of the climb, passing a cairn on the summit to navigate by. I descended to a stone shelter, used to shelter from bad weather, then took a sharp right on a clearly defined trail as I looked down in awe at Loch an Leathaid Bhuain. Over a couple of hours, my world had gone from a fishing lodge by a small road to the wilds of Scotland. There were no buildings visible wherever I turned. I performed a 360-degree sweep and looked on in admiration, appreciating the mix of mountains, rivers and lochs. I could also just make out the coast.

Maldie Burn kept me company as I descended to Loch Glendhu and culminated in a series of lovely waterfalls at the end, as if by sad farewell. A path of limitless Guinness seemed set on staying for the day, and I kept close to the shore of Loch Glendhu, progressing east until I reached the Glendhu bothy.

Dotted all over Scotland, many bothies have interesting pasts, but most were used by farmers to stay overnight whilst tending to their herds. With the advent of improved transport such as quad bikes, farmers can now access their livestock easily, without the need to stay there. Many fell into disrepair until the Mountain Bothies Association came along, founded in 1965. A charity relying on donations and volunteers, the MBA restores and maintains these buildings for everyone to use.

I love them. It's something I can't put my finger on. How excited can a person get just by staying the night in an old stone building in the middle of nowhere? I mean, it isn't up there with a trip to the cinema, a couple of beers and a curry is it?

But these eclectic little shelters dotted in the wilderness, offering a dry and relatively warm night to those in need, have earned a special place in my heart. I've stayed in many, and if my plans entail a trip to Scotland I'll always endeavour to try a new one.

A few years before, on my way to the Isle of Skye for a friend's birthday celebration, was one such occasion. A little research prior to my trip had whittled my choice down to the Peanmeanach bothy near the town of Mallaig because it ticked a few boxes. First, the hike in took three hours, so I figured that might put others off, and it would be quiet. The location was second to none: flanked by mountains and fronting a beach with impressive views across the sea to the isles of Rum and Eigg. It also had great history.

Peanmeanach village once consisted of just a few houses, but the last of the residents left in the 1940s due to the hardships of living in such a remote place. The remnants of these dwellings now lie in tatters, many with just the lower walls still visible; but the bothy, which once served as the post office, has been taken over by the MBA.

As I arrived I was surprised to see a weak column of smoke drifting casually from the bothy chimney, undeterred by a breeze. Being midweek, I thought I might have been alone, but Brett, a chef working in Fort William, had also grabbed the chance to spend a couple of nights there on his days off. He had carried in more than enough supplies for a small army, including some excellent rum and a few succulent steaks. We managed to forage enough wood to get an impressive blaze going in the fireplace. Before long, he slapped two prize slabs of marinated meat on a discarded wire mesh, and we both dribbled at the prospect of impending gratification.

At around 10pm, rum in hand and full in stomach, I strolled outside and took stock of my location. The only sound I could hear was a burn – a small stream – as it ferried water down to the sea, and the land angled gently towards the beach, about 200 yards away. The light was weak and my surroundings barely visible. As I turned around to take in the mountains behind me I could just make out the narrow path to the bothy, which cut a raised, straight line through boggy ground. Something strange met my gaze and, as I struggled to focus, the path appeared to be moving. I walked closer to satisfy my curiosity, and it soon became clear that it wasn't the path that was in motion, but rather what was on it.

As I sat, quietly mesmerised by what I was witnessing, a

line of perhaps 50 deer slowly picked out a route and continued to the beach. They were aware of me, as several paused, looking cautiously in my direction, but appeared untroubled. I learnt later that they frequent the area to lick the salt off the rocks near the beach.

To finish off a memorable visit, in the morning Brett managed to catch a couple of fish. Coupled with a generous haul of mussels, draped in their thousands on the rocks, we gorged on a breakfast fit for kings. It was on that trip that my love for bothies began, and continues to this day.

Chapter 4

Gleann a' Chadha Dheirg

Glendhu bothy looked captivating at the head of Loch Gleann Dubh. It was only 3.30pm, and I'd covered just 12 miles, but I couldn't pass up the opportunity to add another bothy to my tick list. I set my pack down in anticipation; after all, it wasn't as if I had a schedule to keep.

Built in 1880 by the Duke of Westminster, the bothy is a former estate keeper's residence. I saw a house about 100 yards away that appeared to be occupied. It was tempting to knock on the door looking hungry and thirsty, but I figured the residents enjoyed the privacy of their remote house, otherwise they wouldn't be living there. Plus, I imagined they received their fair share of sodden, tired hikers annoying them at regular intervals throughout the year.

The green door creaked hauntingly as I pushed against it and stepped inside, my eyes straining to find a bearing in the dim light. The smell of a recent fire swirled around. As my eyes adjusted, I saw that there was one room to the right and another to the left. A flight of stairs to the rear revealed more space in the loft, but I chose a room downstairs for my stay. A rickety table leant precariously against the wall, bathed in mid-afternoon sunlight streaming in through the window. The bothy logbook, a discarded whisky bottle, and a few tea lights scattered the scratched table top. A solitary chair sat tucked underneath, its slats streaking the dusty floor with lines of shadow.

I still cherish memories from that afternoon. As I returned outside, Scotland was in its prime. The sun soared high, and I felt it burning my arms. A few weak clouds made little impression, and, even though the faintest of breezes struggled to stay focused, the midges were nowhere to be seen. Beinn a' Bhutha lifted skyward behind the bothy, whilst across the loch Beinn Aird da Loch stared back valiantly, attempting to trump its rival.

A burn tumbled and frothed as I collected water, and I sat on a low stone slab outside, leaning back against the wall as countless other guests had done before me. I brewed a cup of Earl Grey and tipped a dehydrated dinner into the rest of the water, then waited for it to come to life. And there I sat, for five hours, reading, squinting at the view and taking in the experience I had come to Scotland to witness, sucking in every last glorious morsel. It was a gentle reminder that, despite yearning to pull in lots of miles each day, sometimes I just needed to rest and experience the place I was travelling through. The god of remaining stationary smiled and gently prodded me with raised eyebrows.

Eventually I returned inside at dusk, succumbing to tired eyes, hypnotised by the last of the light streaming in through the window, painting patterns on the stone walls until darkness tumbled and the sun left Glendhu.

When I woke, the world outside had changed, and it was angry. I wiped condensation from the window and pressed my face against the glass, peering outside. Loch Gleann Dubh was raging, the wind whipping up her waters into a frenzy. Dark clouds had formed overnight, obscuring the summit of Beinn Aird da Loch, and a fierce wind roared. Now the bothy walls screamed in surrender. Rain rattled the roof, and water cascaded down. The ground outside, unable to cope with the torrent, flooded and danced excitedly.

In keeping with the elements outside, my feelings had also changed. I was melancholic. Inspired just a few hours earlier by the high I had felt, I struggled to understand why. Conditions aside, I tried to fathom why my mood had spiralled downwards so violently. I was, after all, still hiking, still in the glorious wild, and still with the prospect of several more weeks doing something I loved. My mind returned to the Continental Divide, and I realised there was a correlation, a state of mind I had experienced during those five weeks in New Mexico. An emotional state that, at times, had reduced me to tears.

Not so much tears of sadness, but more from failing to understand *why* I was sad.

I battled to concentrate, my mind restless and unable to focus for long periods. However, when I could grasp a few minutes of clarity, a brief escape from agitation, I realised with alarm that my sombre moment in Glendhu was familiar. Not just in the present, not just on the CDT the previous year, but clawing with ugly, demonic fingers back into my past even further.

I circled the bothy interior aimlessly for an hour, hands in my pockets, head bowed, and thinking of little. The only activity I indulged in was kicking the floor. Sitting in the chair I stared at nothing, my focus glazed. Despite a lack of desire, I felt no need to bring myself out of it; in fact, despite the complete air of negativity, I continued to indulge in the anguish, confused but strangely content.

Eventually, in despair, I made plans to continue to the next town, rest in a hotel, and get the hell out of Scotland. Nothing made any sense – I wanted to go home.

I tried to check the weather forecast on my phone, but there was no signal. Resigned to a day of being buffeted by the elements, I packed up my gear and said goodbye to Glendhu. I knew Glencoul bothy was just three miles away around the headland and would provide a place to rest if the conditions were as bad as they looked.

I headed east for half a mile before reaching the end of the loch and turning west, back on myself, angling away from the loch shores to begin the climb. The raindrops fell harder and fatter, splashing onto the trail, which had already been transformed into a small stream that gushed over my shoes as I progressed upwards. Angry clouds raced overhead as my path turned again and swung south-east, descending to meet the shore of Loch Glencoul and the bothy.

Beaten into submission, or at least in need of a rest, I virtually fell into the bothy. Water dripped off my poncho, and I wrung out my socks, rubbing my feet to restore some life. I checked the map for options and noted the trail split into two variants from the bothy. Bold red dashes signified the better of the two, but that route headed too far east. A collection of houses known as Inchnadamph was around another ten miles further down the other track. This route appeared more difficult, which I noted from the red dots on the map, but it steered a more direct course to Inchnadamph. I was soaked to the bone and due to get even wetter so decided on the harder of the two trails, comforted by the fact that, hopefully, it would be quicker and there would be lodging at Inchnadamph.

Heading out, I stayed close to the shore of Loch Beag until its waters narrowed into a river. There was no trail, but I knew that my route hugged the eastern side of this river. Hemmed in on one side by a wire fence and on the other by rockfall, I tentatively picked my way across fallen boulders, holding the wire to steady myself. I could see Eas a' Chual Aluinn waterfall, but it wasn't until I neared it that I realised just how amazing it was. The height of the falls stopped me in my tracks. At 658ft, it is higher than Niagara Falls and the highest in the UK. What amazed me was that I had never heard of it, let alone seen it. I sat during a brief respite from the rain and watched the falls plunge over the vertical rock face to feed into the river below. There were no tourists out here; I had the spectacle all to myself. The falls tried to lift my mood, and they succeeded somewhat.

Half a mile further, I had to climb up the side of the ludicrously steep valley. With the storm ripping in from the west, I remained sheltered lower down. However, when I reached the top, all hell broke loose.

I barely managed to hold myself upright as the wind hit. Rain stung my face, and I cowered, turning away. There was only one thing for it; I needed to get inside the Bat Cave.

My poncho protects my body above knee height, and also covers my pack. The sleeves are short, just covering my upper arms, so I can withdraw them inside and duck my head in there, closing the opening. I call it the Bat Cave. It's dry, warm in a clammy sort of way, dim but generous in its proportions. I can eat in there, dry and away from the hell outside, read the map on my phone, and enjoy a few minutes of relative calm to regroup, although it smells musty with undernotes of body odour, so time is limited.

I needed to assess my circumstances. I'd experienced dangerous situations a few times in the outdoors, the incident on the Continental Divide Trail weeks earlier, and a horrific experience during white-out conditions in the Lake District to name just two. I tried to resist the urge to panic, but this onslaught was scaring me. The very storms that I knew slammed the Scottish west coast, the wild elements I had feared, were now getting personal.

The map didn't tally with my surroundings, even though I knew it was correct. I was now in low cloud, hampering visibility, and unable to see the route when I stuck my head out from the Bat Cave. I managed a brief visual of my surroundings before the rain stung my face and I had to turn my head. The ground was a mess, saturated and with no hint of a trail. Wind smashed into me from one direction before turning angrily and knocking me over from another angle a few seconds later. I was pinned against the ropes, this vicious Scottish anger pulling a right hook before ducking and punching from the left. I debated turning back to the bothy but didn't want to waste the effort I had expended getting up the steep valley, so I trusted in my phone's navigation system, took a deep breath, and struggled on.

Ignoring the vague trail, unable to tell whether it was staying for good or not, I did my best to navigate across flat, featureless terrain by linking together a series of small lochs. I slipped through lower ground between Glas Bheinn and Beinn Uidhe, turned east, and hugged the contours to keep my world flat before descending to Loch Fleodach Coire. Finally, at 5pm and after covering just 14 miles in nine hours, I hit tarmac at Inchnadamph. I was exhausted, cold, wet and, to be frank, pissed off. All I wanted was a hot shower, beer, food, a warm bed for the night and a bus back to civilisation.

Inchnadamph Lodge looked more like a grand hotel, framed attractively by a low stone wall and trees. I walked through the gates and up the gently sloping grounds, flanked by lush grass. No-one answered the bell, so I let myself in. Hordes of screaming Japanese school kids were running riot, so when the proprietor informed me the place was booked, I wasn't too concerned – there was another hotel over the road.

I carried on down the drive to the narrow country road and turned left to the hotel. Deer grazed happily in a field, and months later I discovered that the name "Inchnadamph" is derived from the Gaelic *Innis nan Damh,* meaning *meadow of the stags.*

Just a few cars dotted the hotel car park, so I was hopeful they would have vacancies. I entered the lobby trailing a line of water and rang the bell on the counter. The manager appeared and took one sorrowful look at me.

"One night to dry out, sir?" he said with a slight grin.

"Please," I confirmed, smiling weakly to reciprocate the humour.

"I'll put you in room five. Get yourself warmed up and come down to the bar when you're ready. The cook is in residence."

"Thanks very much," I said, holding out a dirt-streaked hand to accept the keys.

I collapsed on the bed, too tired to even get in the shower. Two hours later I woke, my face bathed in a soft, Scottish evening light streaming in the window. Despite the storm having run its course, the forecast for the following day was more rain. I performed a quick situation check: minimal trail food, a dribble of stove fuel, all gear soaked, electronics in need of charging, body sore and state of mind very, very unstable.

I stood under the hot water in the shower, my hands and head pressed against the stark tiles, staring downwards. Dirty water circled below me, eventually running clear. I needed food and a drink but didn't want the company of others in the bar, so I grabbed a magazine on the way to give me an excuse to sit in the corner and wallow in self-pity. I felt no desire to communicate with anyone.

After the second pint of Guinness had slid down, I improved. I was clean, dry, and rested. A decent plate of food helped, and I questioned my decision to leave Scotland.

You're better, mate. Cheer up a little, don't make any drastic decisions you'll regret later. You've been here before. Have more alcohol, you know it makes everything better.

I knew the coach I had originally travelled on to Durness passed right outside the hotel on its way to Lochinver, so I decided to rest and travel to the village the following day to get supplies.

I slept for hours and, walking back over to the Inchnadamph Lodge the following morning in search of cheaper lodging, I wondered if the school kids had vacated. They hadn't, although the owner offered a solution. She led me to a small lodge away from the house, complete with several bedrooms, lounge, bathroom and kitchen. There was a washing machine and room to dry out my gear and take time out.

"It's normally £50 a night," she said. "But, we have no bookings, it's a quiet season, you can have it for £20 if that suits?"

"Perfect, thank you," I replied.

I dumped my laundry in the machine and ran to the road, just managing to flag down the 'luxury' coach that still continued to scrape its way onward. Although the weather forecast had promised little, it was sunny. Just a few unthreatening clouds floated harmlessly.

White-walled buildings lined the street in Lochinver, the village sitting pretty at the head of Loch Inver itself. The loch was still; not even a boat dared launch to disturb the water as I sat, coffee in hand from the mobile sandwich van in the car park. After contemplating my situation on the coach journey, I had decided to continue and endeavour to stop being such a miserable little shit.

I worked through my list, securing stove fuel and restocking on food before taking a seat in the Lochinver Larder, famous for its pies. However, my body demanded a healthier meal, so a large salad and lentil curry was the order of the day. I took one of their cranberry and venison varieties away with me for dinner that evening, though.

I had just one more item on my list, which was proving difficult to source. Since my return from America I had shied away from cigarettes, the consultant's advice still ringing in my ears. However, I hadn't entirely stayed away from nicotine. Back home I'd switched over to an electronic cigarette.

For those not familiar with the e-cig, it comprises a rechargeable battery, a heating element, and a refillable tank, topped up with a fluid usually containing nicotine. Experts in the field consider it far less harmful than traditional cigarettes that burn tobacco and result in a host of nasty chemicals being incinerated and inhaled. The e-cig heating element warms the fluid gently, resulting in a vapour as opposed to smoke. I didn't smoke any more, I 'vaped'.

It seemed to work, as I hadn't touched a cigarette, but there was a small problem, especially in Scotland: I was struggling to find shops that supplied the fluid. Petrol stations and supermarkets stocked it, but the smaller shops in remote Scottish villages didn't. The other issue was that, when I managed to find a shop that held stock, they never had the flavour I preferred.

With the advent of these devices came a whole host of flavoured fluids. I always kept to the natural tobacco flavour, but, for reasons I can't understand, the manufacturers also churned out a plethora of fruit-based extracts and, in some cases, even recipes.

There was a Spar shop in Lochinver, and I entered hoping to replenish my dwindling fluid levels.

"Hi, do you have e-cigarette fluid?" I asked.

"Och, yes my love. Which will you be wanting?"

"Just a natural tobacco flavour please."

"I don't think we have any," she announced, whilst fumbling around the limited stock below the cigarettes. "Any other flavour?"

"What do you have?"

"Let's see now. OK, there's Blue Crush which I think is blueberry. Mediterranean Blast, that's watermelon." She showed me the bottle as if offering verification, like a wine waiter. "What's this, och yes, Real Deal Peach Cobbler and…"

"Peach Cobbler?" I questioned.

"Yes, that's what it says."

"Why would anyone want to smoke a peach cobbler?" I questioned, smirking.

She laughed. "Och, I don't know love."

"It doesn't even taste that good when you eat it, let alone smoke it," I added.

"Actually, the Blue Crush is pretty good," chipped in a customer sifting through the crisp packets muttering something about smoky bacon. "It's very mellow, flavour is surprisingly sweet. It's a decent vape."

Suddenly the shop assistant got very excited.

"Och wait!" she exclaimed. "There is one tobacco left! Traditional Virginia!"

"OK, I'll take the Virginia please, and I'd better have a couple of the Blue Crushes."

I left the shop, deciding to make a concerted effort to cut down on my vaping and preserve the limited stock of tobacco flavour. I love blueberries but, as with Real Deal Peach Cobbler, didn't have the urge to inhale them.

Although e-cigarette fluid was a new addition to my trail shopping list, not being able to procure certain items whilst hiking was a familiar situation. In the States, I always restocked with food and any other items I required at town stops along the trail. In the larger places I could get pretty much anything I needed, but even then, sometimes, I couldn't source everything. I resorted to staying in motels on rest days, searching the internet on my phone for the harder-to-locate essentials. New shoes had to be ordered this way, because my preferred choice of footwear was limited to one model. Certain foods such as my favourite protein powder were handled this way too, as were batteries, which were always far cheaper to order in bulk (I took what I needed and mailed the rest ahead). I would specify a delivery address perhaps a week up the trail and collect them from the post office when I arrived there. I made a mental note to get on the web and order a decent supply of my favourite tobacco e-fluid at some point. It was either that, or I was resigned to puffing my way through various types of fruit desserts.

Some hikers, especially in the States, post all of their food to various stops on the trail so they don't have to resupply in town. Others have spreadsheets with planned mileage, rest stops, and a host of other details. As usual, my planning for the trip had been limited. I don't like long lists covering every possible aspect of an adventure – I prefer to wing it a little. Others plan their trips in fine detail, booking accommodation every night (if it is possible to reach civilisation every day), carrying exact amounts of food, and scheduling arrival times at road intersections where they can reach town for resupply.

My only planning had been the transportation links up to Scotland. I hadn't booked accommodation and, despite Alex's advice, hadn't looked at where I would resupply either. This is how I like my adventures; I prefer the relative uncertainty of not knowing, the enjoyment of a little randomness. If I have a strict schedule, it takes away the enjoyment. If I don't have to be in a certain place at a set time, I am free to revel in a little unpredictability. This approach does get me in trouble, particularly with food resupplies (and e-cig fluid), but I'd rather pull up at day's end at a sweet camp spot, albeit short of planned mileage, knowing I don't have to be somewhere at a defined time. Escaping schedules is one of the reasons I retreat to the outdoors in the first place.

The coach dropped me back at Inchnadamph, and I made my way to the lodge, where two Cape Wrath Trail hikers, Bernadette and Graham, had also checked in. They were hiking south to north, which is the direction most CWT hikers head. As I arranged my laundry to dry near the radiator, they filled me in on their progress to date. The news was expected: long stretches of boggy ground, occasional drier sections, and lots of hills. I confirmed this had been my experience to date also and warned them that the section to the Glendhu bothy entailed a sharp climb with a steep descent down to the valley near Loch Beag.

Resupplied, showered, and laundered, I headed out early in the morning to get a decent start. Joining Bernadette and Graham for a mile, we then split onto our respective north and south bearings as I watched them head upwards, steering a course left towards Glas Bheinn before they disappeared into the clouds. My route south appeared much clearer, and blue skies stretched out invitingly.

With the sun on my face and the prospect for once of a trail of limitless Guinness, for most of the day at least, things looked promising. From the confines of the Bat Cave two days before, I had stripped down to shorts and a T-shirt, smearing a generous dose of sunscreen over my exposed skin.

I followed the River Traligill, gently tumbling from the flank of Conival, reaching skyward to my left. The trail crossed the river and disintegrated into dystopia once more, hugging the lower, steep wall of Conival. The climb and heat of the day made me sweat; perspiration dripped off my nose, and I paused at a small stream to filter more drinking water. Once more, just a few miles out of Inchnadamph, I was back in remote Scotland. Vast swathes of green carpeted the lowlands, changing to more rugged greys as the ground rose to the rocky peaks. Distant lochs glimmered in the midday sun. People had not intruded up here; there were no buildings and no roads. It was precisely how I liked my surroundings.

Squeezing between the lower flanks of Conival and Breabag Tarsuinn, the terrain dipped down and kept distant company with the River Oykel to my right before reaching a firm surface, where I appreciated the increase in my speed.

The Oykel was a beautiful river, and I was glad of the company it offered, albeit for just a few hours. I looked into the crystal-clear water, with just a hint of deep, peaty brown in common with many waterways up here. It wasn't large, either in width or depth, and the flow appeared gentle, in no hurry. Occasionally it stretched out, widening, as if showing off. Anglers love the Oykel – it's famous for Atlantic salmon, and the Oykel Bridge Hotel is a favoured spot to stay for many a fisherman.

After enjoying the flat, smooth fire track through the forest, the old stone Oykel Bridge appeared. It was humped slightly, and lines of grass striped the sides and middle. The Oykel Bridge Hotel, visible just down the road, tempted me over, and I enjoyed a hearty meal, washed down with a fine ale. Although disappointed that I still hadn't located any crappit heid, clapshot, or any other strange-sounding local delicacy, my fuel tank was nevertheless full.

I always seem to walk a good distance after a day's rest. Whether it's my body appreciating the recuperation time, or that I feel guilty about taking a day out and need to make up mileage, I don't know, but today was no exception. Having already walked 23 miles to Oykel Bridge, and feeling strong after the meal, I checked the map and noticed the Schoolhouse bothy was a mere three miles down the trail. Alex had mentioned the bothy, and he had rested there overnight, although he had warned me it was slightly 'creepy'. It seemed the obvious place to stay, and I sped along the track, chased by midges. At least I'd be safe from them inside, and with room to move around.

It was dusk when I arrived, a little sore from the mileage. The bothy sat right by the trail, its corrugated walls and roof reminding me of the dilapidated school buildings I had attended as a small child. I removed my pack and stretched, my back cracking alarmingly, and I walked to the Duag Bridge to fill up with water from the river.

Scotland was still. Dusk tinkered with darkness, the sky still alive with vivid blue and softened by a few grey clouds. Not even a breeze disturbed the silence as I entered the bothy. Tables and chairs dotted one side as if the children had just left, despite it last having been used as a classroom in the 1930s. The other side revealed the usual bothy layout – a raised platform to sleep on and little else.

Another of the simple pleasures of staying in a bothy is spending a few minutes checking out what others had left behind, which was often food. Although the MBA discourage leaving food as it can attract rats, I often wonder whether others left supplies because they were carrying too much or out of kindness to those who may arrive hungry and out of provisions. I like to think it is the latter.

The Schoolhouse didn't disappoint. I found an unopened pack of spaghetti, two cans of tomatoes and some dried oregano – a meal already in the making. A jar of instant coffee, still with the foil intact, a bag of sugar, some vegetable Oxo cubes and a pack of chicken soup completed the larder. I checked the use-by dates (always a wise move in bothies) and, noticing the tinned tomatoes had just a week left to live, helped myself to a can to bulk out my lentils, and saved some for breakfast too.

The place, as usual, had some colourful history. The local authorities had built many of these tiny schools in the remoter areas of the Highlands in the early 1900s. It was common for the teacher to live on site, although this bothy revealed no signs of anywhere to cook or wash, so I could only assume there never were any facilities, or the building had changed over the years. I guessed one side must have been the teacher's accommodation and the other reserved for classes.

The children had to make their own way to school back in those days – difficult enough in Scotland with the weather patterns, especially in winter, but the nearby River Einig hampered their progress even further. The bridge just down the track must have been a pipe dream back when children forded the river. When in flood, the only way to do this was on stilts!

There's something eerie about old, empty schools. They're right up there with abandoned hospitals and mansions, but the Schoolhouse bothy was welcoming and quiet. Ignoring the fear of waking in the middle of the night to the sound of chuckling school kids, I boiled water to brew a cup of tea. I set up my sleeping mat on the raised platform skirting the room, revelling in the space, and spread out my gear so I could locate what I needed easily. I drew the thin curtain, desperately trying to block out light, but it had little effect. I slept fitfully.

Despite my food resupply in Lochinver, I was horrified to discover in the morning that I had neglected to buy coffee and that my supplies had run out. I remembered the jar of instant left on the shelf, but, being the coffee snob that I am, I decided that no coffee at all was preferable to Nescafé. I blamed my low caffeine levels when I took a wrong turn after the bothy and didn't realise until 30 minutes, or a mile and a half, later. Despite making an acceptable start at 8am, it was 9am before I had retraced my steps and started walking on the correct trail. I scribbled in my notepad: *Important – Buy extra coffee, store in different place for emergency rations. Don't forget, you bloody idiot!*

I steered through a shallow valley for most of the morning, the good path I had experienced up to the Schoolhouse having fizzled out. As I looked at the trail ahead, I could just make out a few short sections of firmer footing. My focus centred on the ground a few feet in front. I had to scan the best line, noting irregularities, bumps, and anything that might snag a wayward foot and potentially end my adventure.

Having invariably spent the night drying out my shoes and socks, in the mornings I vigilantly observed the trail a few feet in front, not just for uneven ground, but for water.

The puddles were obvious and easy to steer round, collecting on dips in the trail where water pooled easily. Sometimes these hollows deteriorated to what I can best describe as small ponds. Adamant I would have at least one day on the Cape Wrath Trail with dry feet, I went to great lengths to work around these obstacles – it almost became an obsession.

I sized them up on the approach. Occasionally the shallow water revealed rock or shingle just under the surface where I could, with care, follow a line through. I had to stick to the raised middle of the track, left exposed after vehicles had worn the track down on either side. If the water was just too deep, or long, or both, I had to work my way along the banks of grass and rock either side. One miscalculated move often resulted in a slipped foot back into the pond. During the mornings, sometimes as little as an hour after leaving, I realised that the trail would get the better of me once more as I watched my foot sink into the peaty water after another misplaced step. That morning was no exception, and the quest to preserve dry feet was postponed to the following day, again.

Some days long sections held few memories of the scenery, because my focus centred on eyeing up the best line of attack. I could hike for an hour, or more, and have little recollection of the area I had just traversed except for clipped images. My gaze when hiking focused a few feet ahead, with brief respites to look up and admire my surroundings. I often looked back on my day in the evening, realising with amusement that the sections with the best scenery were often a direct result of the drier and flatter sections of path where I actually had the opportunity to observe the landscape.

Knockdamph bothy came into view at the head of Loch an Daimh, the peak of Cnoc Damh rising behind. I rested on the front doorstep and ate my lunch, once more bemoaning the lack of a coffee shot. Again, nothing stirred. There was no wind, and the sun beat down hard as I tilted my cap down to shield my eyes. Loch an Daimh lay still, elongated, and cupped peacefully at the bottom of the valley.

The trail split and, as usual, the turning I needed headed off down towards the depths of dystopia, whilst the other, an attractive path of limitless Guinness, veered off west towards Ullapool, a popular side trip for hikers on the CWT for resupply. The coach had stopped there on my journey up to Durness, and Alex had also mentioned it. I'd had time to sit with a coffee outside the café and enjoy the picturesque little port, but alas, as my pack was still groaning with supplies, I'd have to be content with memories for now.

I took a sharp left turn at the end of the loch and began what was to be an eventful afternoon. Navigation was the difficult part, even with my phone pinpointing my location. Despite adjusting my course numerous times to stay on the red dots, I never saw the trail all afternoon and instead had to pick out easier ground. Grass up to my knees soaked my legs and shoes, every step sending up clouds of midges. They hovered for hours around my face, trapped in the sweat. I spat them from my mouth, wiped them from my eyes, and switched between wearing my long-sleeve top and my T-shirt, using the former for meagre protection until I became too hot.

The River Rhidorroch veered over and acted as my handrail. Years ago I had taken a navigation course, and the instructor had referred to any feature following a bearing, such as a fence, as a handrail. The Rhidorroch made up for the navigational difficulties, and midge infestation, purely by its sheer beauty – I'd never seen a river like it before. Most of the time it clung to the bottom of a small gorge, the sides of which alternated between about 100ft high to nothing as the trail dipped down to meet the waters. As usual, it was quintessentially Scottish, sparkling clear deepening to dark brown and even shades of bronze as the depth alternated. I tired, my head buzzing with directions, the best line of attack, and negotiating the dips and dives of the trail as it slavishly followed the course of the Rhidorroch.

Mesmerising waterfalls held my attention for as long as I could bear the midges. White, foaming water cascaded ever down, seeking the line of least resistance and plunging into pools of chestnut-brown water before continuing its course into Loch Achall. At the loch's western end, the waters spilled out again to form the Ullapool River.

At times, I held my breath as my route cut right up to the gorge's edge. With clenched stomach I carefully picked my way over slick, wet rock, breathing a sigh of relief as it veered away once more. Rumblings of thunder punctuated the afternoon and became more frequent and louder as the day progressed, until I looked up to see angry clouds directly over me. Suddenly, there were no gaps between the lightning and thunder, and I realised the storm was directly overhead.

For an hour I cowered like a small dog in a fight, not sure what to do. I was away from trees, and I endeavoured to keep to the low ground, but this was easier said than done. The lightning cracks were deafening; I shuddered every time and expected to be reduced to a pile of smouldering ash at any moment. Eventually and mercifully, the wind swept the dark skies east. A grateful landscape began to steam as the sun endeavoured to dry it out, the humidity increase tangible.

I was sad to leave the Rhidorroch, but, as it merged into the River Douchary, the beauty continued. Not thinking my day could get any more wonderful, or eventful, I headed off to thread a path between two mountains, Meall Dubh and Carn Mor.

After hopping over the smaller tributaries of the Douchary, I hauled myself up their steep banks, often on my stomach, my hands flailing around to grasp anything capable of holding me. Graceful it wasn't as I eyed the clump of grass my hand was entwined around, praying the roots would hold. A few rings of tumbled stones, perhaps old crofts or livestock enclosures, acted as bearings, as if the valley weren't clue enough.

Despite a sporadic trail, the ground for once held firm and flat. I took in what was surrounding me, and what my eyes settled on will stay with me for the rest of my life.

I've hiked through the Sierra Nevada, the Cascades, the Pyrenees, the Appalachian Mountains and more, but never had I been so in awe of the wilds that surrounded me as that afternoon in Scotland.

As I progressed, the lower flanks of Seana Bhraigh cleared, revealing the wide gaping valley of Gleann a' Chadha Dheirg. Teasingly, the spectacle unfolded. With each step came a slight change, my angle of view refined, and those lower flanks moved aside just a little more. After all the tweaks, the full majesty of Gleann a' Chadha Dheirg presented itself.

At times, it appeared menacing. Seana Bhraigh, on one side of this valley, rose and stood proud of everything it surveyed. On the other side, of shorter stature but of no less beauty, stood Meall Glac an Ruighe. Both stood guard like sentinels, one minute placid, the next imposing like the gates of hell itself. Steep sides plummeted from a lofty perch, tumbling down from grey to green lowlands, and a ribbon of shimmering water wound down to meet me. The valley walls ran away, stretching further and converging until they curved and carved in to meet each other some 2,200ft up, and two miles distant. As if that weren't enough, Seana

Bhraigh and Meall Glac an Ruighe tore into the sky, attempting to upstage Gleann a' Chadha Dheirg itself.

It appeared the sky wanted in on the action too. The clouds had merged into one huge mass which sported virtually every possible variation of black, white and grey. It rippled and undulated as though alive, breathing, and then, as if I couldn't take any more, the sun appeared again. Through the tiniest break in this veil high above, chinks of light streamed through, flashes raced down to meet me as I appreciated their warmth. The Highlands came alive, light and shadow danced over the lowlands and tore up the mountains. Dark then light, in shadow then illumination, it was the ultimate cinematic experience, and I was the only one there to bear witness.

I sat there, open-mouthed for an hour. No-one shared my corner, nobody experienced what I had seen.

It was all mine.

Chapter 5

The Pit

Hiking through a country is an intimate way to get to know it. Walking speed allows you to experience surroundings at a sedate pace. The landscape appears gradually; mountains and other pleasures appear bit by bit, having been tucked away. Lochs emerge by sheer surprise after cresting a hill, their waters long and thin, nestled between the surrounding slopes. Climbing hills, it's a guessing game when the summit will arrive; but when it does, as if sensing my anticipation, the rewards are often majestic. Peaks stretch away to all points of the compass. Sunlight dazzles, bounces, and casts cloud shadows that race over the world. Why rush?

Progressing at three miles per hour amplifies the travel experience, hones it, and your world unfolds slowly. When progress is this relaxed, we observe the details, and aspects of travel we would otherwise miss show themselves.

For example, I soon learnt to appreciate how Scotland mopped up after the rain. Countless trickles, streams, lochs, and rivers all play their part in allowing water to escape to the ocean. It was everywhere, an infinite flow seeking the path of least resistance. The landscape becomes fractured over millennia as moisture erodes, carves, and chisels away at the surface. The results are magnificent.

I can hear this journey, unseen but trickling beneath me somewhere. Meeting burns, it builds in intensity before spilling into the rivers – the real arteries. When I look up at the rock faces after a storm, streams of water cascade down, streaking the stone like tears.

The lochs scoop up these sources and, for a time, hold on to their bounty before releasing it to another river outlet. On windy days, not uncommon up here, white horses charge across the surface. On calmer days, walking beside these mirrored wonders, the merest hint of a wave laps quietly against the shore.

But it's the rivers that have caught me by surprise. So many of them, cascading through valleys, racing towards the coast. The noise is deafening, crashing and foaming, roaring over jagged rocks. They have carved valleys, gorges, and entire landscapes.

The true gift of this hike so far is the waterfalls. I understand now that these glorious displays of power are common in a landscape such as north-west Scotland, but it wasn't a sight I had prepared myself for.

Often, the water is tinted dark brown, picked up from the peaty soil, but it's gloriously clear. Angry rivers are stunning. Ivory sheets of water tumble over precipices and appear to pause, plummeting to waiting pools. As I leave, the roaring subsides, and the river continues.

The gorges amaze me. I peer over the edge, stomach clenched, and hear the water boom beneath, seeing these torrents stopping at nothing on their journey.

Scotland is renowned for being wet, yes. I could be polite and temper the argument, but it's true, this corner of Great Britain gets a lot of rain. But, having now walked through a good part of it, I have a newfound respect for the beauty of this journey of water. If it weren't for the storms, these raging weather patterns I often cursed, then Scotland wouldn't be the place it is today.

Reluctantly, I left Gleann a' Chadha Dheirg. I could have stayed there for days, but, as with all hikes, forward motion, the pursuit of progress, and the goal of finishing propelled me onwards.

Stumbling over uneven ground I descended towards the few isolated and scattered houses known as Inverlael. A forest provided welcome shade from the afternoon sun, which I had no way of hiding from in Scotland's openness, and I sped quickly down the fire track switchbacks. The River Lael gurgled beneath as I crossed over the bridge and arrived by the A835.

Not only was I exhausted mentally, and struggling to cope with the psychological minefield that constantly affected my mood, but physically I was suffering as well. I had two impressive blisters, my legs needed dragging along for the ride, and every sinew stretched like overtuned guitar strings, seemingly ready to snap.

Despite the high I experience when hiking long distances, both body and mind take a battering. The constant repetition, the continual pattern of placing one foot in front of the other has consequences. That area inside your shoe that seems slightly uncomfortable on a three-mile stroll is amplified tenfold over the course of a long day. The hip belt rubbing on a training hike suddenly becomes unbearable by mid-afternoon.

However, these were more than the usual niggles. My body was lethargic, drained, and listless. My mind, once more, was playing games. I had started to score my mental state – it ranged from a 1, for as bad as it could get, up to 10, where I was elated just to be alive.

This was strange ground for me, unknown territory. I'd had mood swings in the past which were short-lived. I'd never been low for longer than a day or two before. But now, once there, in that deep pit, it seemed an absolute impossibility to climb out.

It's dim down there, cold, wet, and gloomy. Peering up I see daylight, but it's too far above me, and I can't reach it. I'm a prisoner in a vertical tunnel. The walls are moist, slippery rock offering no grip. I feel the moss, the green slime, but can only just make out its weak glint. I hear nothing except my breathing and the occasional drop of water, echoing around my prison. As I exhale, my breath condenses and rises. I watch, envious, for it is the only part of me capable of liberation.

A pit, that's where I am. One deep, unbearable shaft.

My head, intent on inflicting more pain and confusion, appeared to be rearming with stronger weapons. As if it knew that I was aware of this dangerous game, and that I was privy to its plans, it upped the stakes. Now I felt depressed, the low mood was lasting longer, and it was an immense battle to scale the pit walls to freedom.

Sometimes I never even attempted to escape, resigned to my fate. Other times the pit wasn't so deep, the rim closer and the walls drier. Depending on my inclination, I climbed out or reached halfway before I fell again.

But the worst part was not knowing. Waking up every

morning, I cracked open one eye and peered out into my world – a place where I didn't know how my mood would be for that day. I had no control. I had to accept what was served. It was this fear I dreaded the most, and the pit was always primed to devour me.

A mere seven days and 100 miles into the trail, and I was exhausted. Two signs by the road offered accommodation, but also displayed 'no vacancy'. Carrying on, I crossed a bridge over the River Broom before pausing by a small road junction. Another sign advertising rooms, bereft of any 'no vacancy' note, swung from a post in the light wind. The road stretched dead straight through fields dotted with sheep. Low stone walls ran parallel, and buttercups splattered yellow through the verges, waving as I passed.

Following the path for a mile before finding the bed and breakfast, I paused at the gate, wondering if 9pm was too late to be asking for somewhere to stay. I rang the bell, and a man opened the door.

"Hi, sorry it's so late, but I was wondering if you have a room for the night?"

"I'm not sure," he replied. "My wife, Mary, runs that side of things, but come in, I'll phone her."

He led me into the kitchen and offered a seat before calling his wife.

"Yes, we have a bed for you," he said, replacing the receiver. "Do you want some food? We cook an evening meal for the guests but they've already eaten. I've got takeout curry we can't finish. I can reheat it for you?"

"That would be fantastic," I said.

Mary arrived as I was finishing my meal and showed me

to my room. After showering, I struggled to show any interest in a documentary on the TV and drifted off.

I came to in the morning after 12 straight hours of sleep. I never sleep for that long, which suggested something was amiss physically, but, feeling better, I didn't question it. Mary prepared a hearty breakfast, and I geared up, better for the rest and sustenance.

A steep climb greeted me out of Inverlael, culminating at Loch an Tiompain. Not relishing the climb at the start of the day, I studied my phone for other route variants to meet up with the CWT later. There seemed to be a solution. A stony track, running parallel to the River Broom, worked its way past the Inverbroom Lodge to a bridge over the river. It was around five miles away. From there I'd nip up a steep bank, cross the A832 and follow the road for a couple of miles before disappearing off into the mountains. After handrailing Loch a' Bhraoin, a faint trail bisected the CWT again. I decided to take this alternative.

It was flat, easy walking. I nodded to a few estate labourers repairing the road, passed the lodge, and, just shy of two hours later, arrived at the bridge. A twisted mess of iron plummeted to the river below, while a few sections of rotten timber clung on helplessly. The gorge was too steep to even contemplate climbing into. It was a dead end.

I tried to remain upbeat but felt myself teetering on the edge of the pit once more.

Two choices, Fozzie. You either let yourself fall, or you fight.

I stayed on the surface, barely. When I returned to the bed and breakfast, Mary looked surprised to see me.

"That was quick!" she exclaimed. "You OK? You don't look good."

"I'm fine," I replied weakly, lying. "Don't suppose you have a room for tonight again do you?"

"No, we're fully booked tonight, have a group of cyclists. But wait, I know someone down the road."

She spent a minute on the phone, turning in my direction and smiling positively.

"Iain at the Forest Way Bunkhouse has space. It's four miles away. Come, I'll give you a ride, I need to see him anyway."

I had passed the bunkhouse earlier; it sat on the other side of the river. Iain welcomed me and showed me around the simple, wood-clad building. There was somewhere to cook, laundry facilities, and Wi-Fi, and it seemed a relaxing place to hang out.

"Do you need anything?" Iain asked.

"I could do with a glass of wine."

He laughed. "I usually get requests for pasta or a can of tomatoes! I keep a few bottles for the B&B guests, you're welcome to one of those. White or red?"

"White. Thank you."

I spend early evenings at home in the kitchen, where I chill out. I'll happily spend an hour preparing a meal, in between reading the paper and sipping on wine. People watch the TV or go out; I enjoy cooking, which is precisely what I needed.

I bartered a deal with another hiker called Bob for some of his pasta in return for checking his email on my phone, as his was dead. There was leftover pesto in the fridge, and parmesan, both of which Iain confirmed were up for grabs, and someone had even left a newspaper.

Fed, inebriated, and relaxed, I turned in, determined to get a grip on whatever issues my psyche was throwing at me.

I completed the two-mile road walk back to the CWT in the morning. The climb to Loch an Tiompain proved easier than expected, and the quick gain in altitude rewarded me with stunning views back down to Inverlael, where I picked out my route from two days before. The bed and breakfast, the bunkhouse, the long, stony track I had walked down, and even the broken bridge were all visible.

I had five miles to Corrie Hallie, where I hoped I might find a coffee. The splendour of Gleann a' Chadha Dheirg still played on my mind from two days earlier, but the view that greeted me on the descent to Corrie Hallie made a valiant attempt to upstage it.

Bidein a' Ghlas Thuill, part of the An Teallach range, considered by many to be Scotland's finest mountain, roared up from the depths of the Highlands to dominate the landscape. Its lesser satellite peaks (in elevation but not magnificence) of Sail Liath and Glas Mheall Liath stood guard like rock soldiers either side. It just stopped me dead in my tracks. I smiled; for every time I had experienced a harsh day psychologically, Scotland's wilds attempted to distract me the day after. This glorious land of wild mountains was proving the ultimate counsellor.

I reached the Dundonnell River, lovingly shaded by a clump of trees, and smiled as children played excitedly with their dog in the water. There was one shop in Corrie Hallie selling gifts, and my tentative enquiries into coffee availability were met with confused looks by one of the staff. I took her expression as a no.

It was a glorious day and, in hindsight, one of the best of my entire traverse of Scotland. The weather was perfect; sunshine bathed my world as mesmeric ivory clouds hung motionless over a blue canvas. The scenery showcased itself, not just an amateur production, but more the full Broadway spectacle.

I crossed the road and followed the route, just above the Allt Gleann Chaorachain river, and climbed until I reached a split in the trail. The right fork wound to the Shenavall bothy, commanding an epic view at the head of Loch na Sealga, but, as attractive as its location appeared, it was still morning and too early to consider a stop for the night.

I paused, sitting to rest and eat. Three mountain bikers stopped and enquired about my hike. Graham, who lived nearby, knew the trail well and, after pointing out the bothy, motioned south where the trail continued. A wide valley swept up gently with Beinn a' Chlaidheimh to the west and Creag Rainich to the east. Abhainn Loch an Nid snaked through the scene with ease, glinting. He said that Loch an Nid was around an hour away, adding it was his favourite loch and a sight to behold. They wished me well and careered off down the hill, spraying gravel in their wake.

I ventured into the valley, feeling I had entered a dramatic gateway. Looking east, rock soared steeply up, whilst over the other side of the valley the incline, no less dramatic, rose gently. The rock faces streaked with cascading streams of water, and I paused, stopped my breathing, and cupped my ears to listen intently, just able to pick out the sound of gushing water.

Reaching Loch an Nid, I immediately related to Graham's praise. The calm waters culminated at the southern end with two fishermen trying their luck, but they shook their heads and shrugged shoulders when I called over, enquiring if they had caught anything.

Steering a course south-west between Sgurr Dubh and Beinn Bheag, I climbed on a bearing towards Lochan Fada, where I planned to camp overnight.

I spotted four other hikers ahead and gained on them, negotiating uneven ground, hopping over streams and veering around rougher terrain. I caught them quickly, noting their packs, which appeared huge, and guessed they were out for a few days. We stopped and rested together.

"Are you with the Cape Wrath Trail?" one asked in broken English.

"Yes. You too?"

"Yes, but we move very slow. It is hard! The hills, the ground is rough. And the weather!"

"Where are you from?" I asked. Only one of them spoke as the others smiled and nodded. I assumed he was the only one who spoke English well enough to communicate.

"The Czech Republic. We come here for two weeks to see Scotland. It is very nice. Do you, sorry, have you been to my country?"

"No, but it's on the list."

They left before me, but I soon caught up and passed them, wishing them well.

I'd seen seven other people in one day! Despite my yearning for solitude before the trip, this made a welcome change, but I needed to talk to the government department responsible for calculating the population figures. After all, their statistics were way out. Today I'd seen seven others, the most in one day so far, but still way, way off the 171 per square mile they were laying claim to!

A light rain fell as I adjusted my course to cut the corner of Loch Meallan an Fhudair on my descent to Lochan Fada, now visible, stretching and twinkling in the early evening light. After a rough descent, I arrived and set my pack down, sitting to rest on the pebble beach.

The rain stopped, and the skies appeared to clear. I found a narrow section of flat grass bordering the shore to erect my tent, which utilised one of my trekking poles as the main support – a weight-saving feature that meant I didn't need to carry a dedicated tent pole. I couldn't find my trekking poles and realised I had lost them. Most of the time I was never without my poles, but occasionally I tucked them in the side of my pack and walked without them. They weren't there, and I had no idea where they had gone. Despite scanning the surrounding area for a length of wood, or similar support, I knew there was no way I could pitch my shelter. Nine days into my hike, and it had rained every one of those nights so far. The situation didn't look promising.

I looked skyward, in part to check the weather, and in part hoping for salvation. Threatening clouds glanced the summits but were moving away, and the evening looked calm, although I had no way of gauging what might happen later that night. I had no choice but to spread out my groundsheet, sleeping mat and bag on the pebbles. After eating, I spent an hour staring across the length of Lochan Fada, digesting a wonderful day and praying for a dry night. Complete darkness didn't arrive until just before midnight.

Despite the wind increasing in the early hours and inflating my sleeping bag like a windsock, Scotland remained dry. Although tired, I slept little, constantly waking, almost out of choice, to take in the stars and the morning light. I woke with a checklist containing two very important items: a new pair of trekking poles and a battery pack, which I had also lost. My phone had just enough juice to last me until Kinlochewe, 10 miles distant, where I hoped I could get transport to a larger town with an outdoor store. I was starving and needed sustenance. My appetite was ferocious; despite my recent resupply, my food bag had dwindled to a few oatcakes and a little dried mango.

Selecting the right food when I hike is always a battle. I struggle to get the right mix of savoury and sweet, ending up with too much of the one I don't want and not enough of the other.

I also have wild cravings. For example, I don't have a

sweet tooth. I rarely eat dessert and hardly ever buy chocolate bars or confectionary unless I'm hiking, when I crave sugar. I'm interested in nutrition, and my normal diet is very healthy – plenty of raw vegetables and salad, fruit, nuts, and good fats such as olive and canola oil. I'll go for days without meat, preferring fish, and try to get my protein from sources other than animal flesh – although I enjoy steak and bacon as much as anyone.

My hiking food bag has a capacity of 16 litres, and I arrange various items in Ziploc bags. Tea, coffee, powdered milk, and breakfast granola account for one. Evening meals, mostly flavoured rice dishes, make up another, along with dehydrated vegetables, which I add to the rice to liven things up. My 'healthy' snack bag contains stuff such as almonds, cashews, dates, chia seeds, and a couple of varieties of dried fruit. I reserve one of the smaller bags for spices and herbs – especially curry powder, which has saved many a meal from mediocrity – and rumour has it that soy sauce and Tabasco will always lurk somewhere.

My 'sweet-tooth' bag, as I refer to it, is crammed full of processed, high-sugar-content items I'd never go near at home, but I believe that, as poor as the choices are when hiking, what my body craves is what it requires. Plain chocolate I love, especially with a high cocoa content. That's where the healthy content of the sweet-tooth bag finishes. The rest is filled with Skittles, Minstrels, Snickers and Double Deckers. I never eat these foods off trail; in fact, I dislike them, but once hiking my body demands energy, and sugar is the preferred hit. Maple syrup has wormed its way into the equation, although it is extremely good for you. I use it as a sweetener for my tea and coffee, and my granola always gets a generous squirt of the deep brown treacle. If I can't get maple on the trail, I'll settle for honey. Phillip Colelli, with whom I hiked much of the Appalachian Trail, always had a squeeze bottle of honey with him and swore by it. When he tired or just needed an energy hit, he'd fumble around in the side pocket of his pack and pull out the sweet stuff, flip the cap, and load a healthy squirt into his mouth. He said he felt the effect almost immediately.

My other weakness on a hike is fat. My preferred, healthy choices at home are difficult to hike with because they weigh too much, and I go through a lot of good fat, so they never last. Oils are also notoriously difficult to transport. Somehow, despite even the most secure of containers, they manage to leak. By the time I reach town, my body is screaming for the bad forms of fat, such as meat.

I arrive in civilisation obsessing about breakfast. I can smell it, and my inbuilt bacon beacon steers me straight to the local café where I sit, bolt upright with hands firmly grasping cutlery, only releasing them to pour black coffee down my throat, waiting for my fat hit.

On the big hikes I've completed in America, it was normal to go through 10,000 calories during a day in town. The primary goal was to counteract the calorie deficit incurred during the previous days' hiking. I devoured bacon, eggs, beef, cheese, full-fat milk, fried foods, and ice cream in far greater amounts than I'd consume off trail. And I never struggled to put them away. After each sitting, often four huge servings each day, I was constantly thinking about the next meal. Despite the obscene amount of food most thru-hikers put away in town, we still struggle to maintain body weight, often losing some. There's a quote I read in America, I forget the source, but it sums up a thru-hiker's appetite perfectly:

I've never seen skinny people eat so much.

81

A mere two miles on from Lochan Fada, the heavens opened. I stopped quickly, grabbed my umbrella, and retreated to the Bat Cave. The wind buffeted, shoving me rudely off the trail as I slavishly followed the course of the Abhainn Gleann na Muice river down towards Kinlochewe.

An abandoned outhouse provided a rest stop, sheltered from the pounding rain, where an inspection of my supplies revealed that my tobacco-flavoured e-fluid was running out and a switch to the blueberry variety was imminent. I checked the map, which revealed a few inviting symbols in and around Kinlochewe. A blue beer tankard suggesting alcohol lurked somewhere, along with a cup icon indicating caffeine. There was a shop symbol, and a bed promised accommodation. I knew Kinlochewe was a small village, with little hope of supplying trekking poles or a battery pack, but at least a place to eat and dry out looked promising, and I could make enquiries about onward transport to a larger town.

I spilled out, somewhat sodden, onto the A832 and followed a path by the side of the road to the village. The white exterior of the Kinlochewe Hotel beckoned me over, and I emerged from the Bat Cave, sweaty from the humidity. I was filthy and attempted to make myself look presentable before entering the bar, quickly checking my reflection in the window. I ran my hands through my hair, which refused to budge from bolt upright, wiped some dirt streaks off my left cheek, and hoped for the best.

My bacon beacon was fluctuating wildly in anticipation of a cooked breakfast, but, alas, the bar couldn't deliver. The barmaid smiled at my appearance, no doubt used to weary hikers escaping the wrath of the weather, and I watched eagerly as she adjusted the coffee machine to deliver my Americano just how I like it – black, with an extra shot and half full.

"It's not great weather out there, love, is it?" she offered, carefully placing my caffeine hit on the bar in front of me.

"You could say that," I replied. "Do you know the forecast?"

"Och no. I don't check it anymore because I usually end up disappointed!"

"I know what you mean. So, breakfast is finished, but are you doing lunch?"

"Yes, of course," she replied, handing me a menu.

I was hankering after any form of carbohydrate, a natural yearning on a hike. Chips seemed the obvious choice, but another option appeared far more attractive.

"Haggis, neeps and tatties please," I requested, feeling strangely chuffed that, being an Englishman, I was allowed access to the legendary dish.

"Take a seat, love, and I'll bring it over."

I sipped my coffee and took stock. Priority was to get to a gear shop for my trekking poles and battery pack. There was nothing in Kinlochewe, and, being Sunday, any form of onward transport was unlikely. The nearest train station at Achnasheen was eight miles away, and I wondered if there was a local taxi.

I checked the map while shovelling haggis into my mouth as if my body were due to cease all operations until I, also, had replenished my batteries.

"Can I get a train to Inverness from Achnasheen?" I asked the barmaid.

"Aye, you can but the last train was at three," she said with a slight shrug of the shoulders, glancing over her shoulder at the clock. "Are you hiking the Cape Wrath Trail by any chance?"

"I am yes."

"Well, the next train is in the morning, but the taxi will

cost you fifteen pounds, if you can even find one. There's a train station at Strathcarron, twenty miles along the trail, that will take you straight to Inverness. Do you need somewhere to stay tonight and dry out?"

"Yes, but the hotel looks expensive."

"Aye, but the bunkhouse has beds and it's cheap, ask in the hotel."

"Thanks very much. Great haggis by the way."

The hotel was expensive, and full, but I secured a bottom bed in the bunkhouse. I hate the top beds, vertigo notwithstanding – I never sleep up there. The other problem with bunkhouses is what I refer to as the 'unpleasantness factor', or UF. Various factors decide the score, ranging from a 1 (very poor) to a 10 (perfect). For example, the number of occupants is a sure-fire way for the score to slide downwards dramatically. More people equals less space, especially as it's not just bodies interfering with my world, but luggage (namely backpacks) as well. In my younger years I dealt with it, in fact I liked it because it was exciting – all these other travellers from around the world with stories to tell, living the nomadic life.

You wouldn't imagine that poor weather could affect the UF, but it does. If everyone comes in soaked to the bone, then the interior transforms into a Nordic sauna. Windows steam up, moisture drips down the wall, and, if someone calls you from the other side of the room, finding them entails waving your outstretched arms around whilst trying to find a way through fog that wouldn't be out of place in a Stephen King horror movie. Couple that with everyone trying to grab the one, solitary shower, and the subsequent additional clouds of steam, and you half expect to reach the other side of the room and see someone throwing water on a pile of hot coals.

The culmination of a poor bunkhouse UF peaks after around 10pm. Everyone's been out to eat and, being on holiday, to drink. I'm tucked in bed by then, so I must endure the returning crowd. We know that inebriation results in chatter; someone at some point will fall over (followed by chortles from the rest), and it takes forever for them to calm down and shut up.

The final annoyance is that, with full stomachs from rich food that people don't normally consume (such as haggis and Guinness), flatulence is rife. Farts of all shapes and sizes reverberate, echo, and ping off the bunkhouse interior, followed by the inevitable aroma of digesting Chardonnay and scallops. When people have finished breaking wind, around 5am, it's just about time for those making an early start to get out of bed and pick up the conversations (and farts) from where they left them. Once more the room descends into fog, as those who missed the shower the previous evening (because they couldn't find it through the steam) grab their opportunity. Clanks of pots and cutlery float in from the kitchen, and every 10 minutes someone's alarm clock kicks in.

As it turned out, Kinlochewe bunkhouse wasn't too bad, about a six on the unpleasantness scale, and I slept reasonably well.

During one of my hikes on El Camino de Santiago in Spain, I spent the night at a bunkhouse in the charming village of Samos. The evening began promisingly as the caretaker advised that one room was full and that I would be the first guest to stay in a new, empty room.

I picked a sweet spot by the window, so I could enjoy some fresh air overnight, and it also afforded extra space around my bed for my belongings. I went out, enjoyed a great meal, and returned to find six of the other beds had been occupied by cycle tourers, judging by the panniers strewn haphazardly around the room. Wet clothing hung everywhere, dripping on the floor and reducing the air quality to a mix of body odour, road dirt, and soiled chamois.

Manoeuvring gingerly around sodden clothing towards my bed, I slipped on the wet floor and landed painfully on my backside. After checking nothing was broken, I moved some of my belongings away from more dripping clothing, got into bed, and drifted off.

Three hours later, at 1.30am, I was woken by six drunk Italian Tour de France wannabees stumbling aimlessly around the room in the dark trying to locate their beds. Childish giggling ensued, one fell over, and another smashed his head on the upper bunk to cries of laughter from the rest. Then they started to chat constantly as if they had just woken and were making inroads into their second breakfast espresso.

I felt I had been more than patient, but, after the third fart, I lost it.

"I hate to interrupt your 2am conversation but would you all mind shutting the fuck up please!"

The silence was golden.

Chapter 6

Manicured Poodles

In the morning, I took stock of my situation while trying to locate an elusive trailhead somewhere along the A832. It was day 11, and I'd covered 137 miles, an average of 12.5 miles a day. This didn't concern me too much – I often average around 15 miles a day during the first week to bed myself in, and I wasn't in any hurry on this hike. I still hadn't dried out. My shoes had been permanently wet since day one. Physically I was best described as reasonable, although on several days I had started out well, only for my muscles to rebel mid-afternoon. Again, I wasn't worried, because some of my hiking fitness had left me during the period of inactivity since returning from the US.

But I was uneasy about my psychological state. It scared me – not just the huge dips in my mood, more the actual fear of deteriorating. It felt as though the pit were lying ahead, covered in undergrowth, hidden, just waiting for me in a moment of weakness.

I couldn't fathom the reasons behind my sadness. I was

hiking, roaming through Scotland, as free as I could ever hope to be, and revelling in my one big love: solitude. It made no sense. I was living the life I loved; I should be elated, full of expectation. However, the fear was a constant presence, bearing down, belittling me, toying, teasing, and playing its immature games. Sometimes it seemed even thinking about the pit was enough to topple me.

Slavishly sticking to a bearing, I aimed for a derelict croft, squelching through sodden fields, climbing through barbed-wire fences, under overhanging foliage, down rocky banks, and across streams. Finally, a sign for the trail appeared and beckoned me upwards and away from the A' Ghairbhe River. Skirting the lower eastern flank of Carn Dhomhnuill Mhic a' Ghobha, a forest engulfed me. I stumbled over rough forest roads, sliding on slick mud while trying to find the narrow trail which plummeted down to intersect the A890.

I stopped around midday, removed my pack, and rummaged around in my food bag for something to eat. A few midges hovered around my face. Although elated at having found some long-lost oat cakes, I glanced at my arms in horror. My forearms, and the rest of my body, were absolutely covered in midges. I ran down the track flapping my outstretched arms while trying to keep hold of a precious oat cake. To an onlooker, I must have resembled a child pretending to be a plane taxiing down the runway. I swiped at my face and spat out a small colony, vigorously rubbing my hair. Running back to my pack, I grabbed it and sped off, undecided whether to hoist it in mid-flight or brush the midges off it.

Stumbling through more overgrown land and waist-high bracken, and with little sign of any trail, I happened upon another abandoned building. This was becoming a regular habit. These forgotten crofts, homes, and decaying farms looked like nothing more than toppled walls. They made useful places to rest; the stones provided a dry seat off the damp ground. Now quiet and forlorn, they hide a history of darker times in Scotland known as the Highland Clearances.

During the 18th and 19th centuries, tens of thousands of people were forcibly evicted from their land by aristocratic landowners who deemed their small-scale agriculture unprofitable. The aristocracy wanted to switch to sheep farming, which was far more lucrative.

The clearances took over a hundred years. Those unwillingly evicted lost their land and were driven to take up new lives anywhere they could. Many fled to America, New Zealand, Australia, and Canada. Others travelled south to England to take advantage of the employment possibilities offered by the Industrial Revolution. Some moved to towns such as Ullapool, Inveraray, and Plockton which had been founded to take them in, or to the bigger cities. Under Scottish law these tenants had no legal protection and, despite their understandable refusal, were forcefully removed.

There are other possible explanations for the clearances, with the dark shadow of politics never too far away. Some say the British establishment encouraged the act, keen to be rid of the archaic Scottish clans responsible for several uprisings against the system.

Even the weather played a part, or at least fuelled the process, causing several poor harvests. A rising population exacerbated the situation, and many saw the clearances as a chance to find a better life overseas, clamouring to board vessels bound for the Americas.

Today these abandoned buildings dot Scotland, from solitary ruins once housing a few families to others big enough for several hundred. The roofs were often thatched from local materials: bracken, grass, and rushes. These materials degraded to leave just tumbling stone walls, a haunting reminder of days gone by.

Even in the vast wild areas of Scotland, in the middle of nowhere, miles from a track or road, I'd stumble upon the clear outline of a place someone once called home. It must have been a hard life, especially during the winter, but it was a life, a living, and provided for those who chose to stay there.

Reaching the road, and glad of level ground, I walked on the verge until crossing the railway line at Craig – nothing but a few scattered houses. A bridge over the River Carron continued onto a rough track, dipping in and out of woodland where I struggled to find any trail. Squeezed by the river on my right and steep hills to my left, I had little choice but to relent as I was funnelled down towards what was to be one of the hardest sections of the entire hike.

Loch Dughaill stretched ahead for three miles. I checked the map, which showed a clear trail along the southern bank. Wandering by the last section of the river before it entered the loch, I picked a way through grass and rushes, my feet sinking in to the sand, before emerging at the head of the loch. I'd had no quarrels with the map until this point, and it wasn't even the map's fault, but the suggested line of attack along this southern edge beggared belief.

At first it was just a case of negotiating uneven ground strewn with large boulders, many slippery from the water and weeds. However, as the camber steepened, the path disappeared to nothing, and I had to tackle a sheer bank of rock perhaps 20 feet above the water. With the relative comfort of at least knowing I had water to land in should I fall, I clung precariously to the rock, forcing myself and my pack around overhanging branches. My feet slipped on the damp rubble, and my hands constantly searched, flailing, to secure a confident grip. After what seemed like hours, I reached the end of the rocks and the sandy bank of Loch Dughaill once more, collapsing, exhausted, and glad to have suffered nothing more than a few scrapes and a bloodied knee. I looked back at the trail, and my map, in disbelief.

Scottish trails are different animals from my usual routes in West Sussex. Back home, if I took a footpath shown on my map, I'd be assured that, firstly, it would be there, and secondly, it would be in reasonable condition. Up here over the border, I found myself doubting not only the map but my sanity. I began to mistrust my smartphone – perhaps it was malfunctioning? Maybe the cross-hair icon was half a mile out and I should have been treading a path down the other side of the loch, or over in the next valley?

I wondered whether the cartographer, or whoever mapped the CWT in the first place, had decided to put me through hell and back. Perhaps they'd lost a bet down the pub and plotted parts of the route through the most inhospitable terrain known? I could hear them, bent over their whiskies and pointing at me, chuckling uncontrollably at my misfortune.

Despite the humour attempting to placate me, I reacted to the dangerous section of trail badly. Instead of remaining calm and letting my anger slide, my frustration boiled over, and I began shouting.

"Fuck this shit! What the hell was that all about!"

Slamming my pack on the sand, I tore off my jacket and threw it down, then kicked my pack.

"This is fucking ridiculous! I'm not even enjoying the bloody hike! Screw Scotland! I hate the place!" I screamed.

Eventually, my anger abated and I sat on the grass, my heavy breathing slowly calming. I could feel my face screwed up in anger and frustration. Nothing seemed to be going right for me this year. My Continental Divide Trail had ended abruptly, and I hadn't enjoyed it. To reward myself for that failure, I thought the only solution was to embark on another hike in a country that I knew was beautiful but was sure to test me even more.

I shook my head in disappointment, rubbed my eyes, and cursed for even making plans to hike across Scotland.

As if sensing my discontent, the forces that be rewarded me on the approach to Strathcarron, and I enjoyed a leisurely pace all the way on a minor road next to the railway line. The village itself consisted of little more than a few houses, the railway station, and a hotel-cum-bar. For once, I had timed my arrival well and had just an hour to catch the last of only four trains to Inverness.

In the space of that hour, I put back three pints of Guinness and a double scotch in the bar. I knew alcohol was not a solution to my woes, but it provided temporary relief. I caught the train and bumped along en route to the big city.

It felt strange to be back in Inverness so soon after my stop on the bus to Cape Wrath 12 days earlier. The city was peaceful; in the early evening the commuters had already made their journeys back home, and a calm had descended. A soft, low summer light played with shadows around me, and barely a breeze disturbed the quiet. I decided against a taxi and strolled through the streets to find the same hostel I had stayed in a few days earlier.

There were no coaches in the car park – always a good omen at a hostel – and inside the only signs of movement came from a family sat around the reception area. I checked in and showered while watching the brown murk of the Cape Wrath Trail spiral down the plughole.

With my drinks at Strathcarron wearing thin, and eager for more to numb my feelings, I sat at the bar in the Chieftain Hotel eating fish and chips and indulging in more drink. After 12 days on trail, and 157 miles under my belt, I felt bruised. Physically I hurt, but I could always handle the pain my body threw at me. Not only was my psychological state continuing to deteriorate, my resolve to battle it was weakening.

I don't do cities – too much noise, people everywhere, the air smells funny, it all seems rushed and impatient. However, Inverness is a worthy exception. The River Ness glides past with minimal effort, as if whispering, careful not to disturb anyone. Impressively broad but shallow, it breaks up the city into a wide ribbon of calm, the grassy banks gently shelving and speckled with the occasional tree. Even the residents seem to make that extra effort to keep their house fronts tidy, with flowers poking up over stone walls and freshly painted frontages.

Beneath the calm surface of Inverness lies a brutal and bloodied history. Due to its strategic location on the River Ness, and the early ferries that carried goods over the water, it developed as a focal point for armies to access the Black Isle from North Kessock on the south.

The Craig Phadrig hill fort was occupied from the 4th century BC when the area formed the capital of the Pictish kingdom. King Brude of the Picts had a fortress here as far back as the 6th century.

In 1040, Macbeth reputedly murdered King Duncan at his castle, which has since been destroyed, and these events were made immortal by William Shakespeare's *Macbeth*, although it is thought the play is highly fictionalised.

Perhaps the most famous battle took place on Drummossie Moor, near Inverness, in 1746. The Battle of Culloden saw the Jacobite forces of Charles Edward Stuart decisively defeated by loyalist troops led by William Augustus, the Duke of Cumberland.

Stuart's choice of rough, boggy ground proved catastrophic for manoeuvres, and his meagre arsenal of weapons, comprising just swords and daggers, proved no match for the cannons and guns of Augustus's army. It was all over in an hour, with between 1,500 and 2,000 Jacobites killed. Augustus suffered fewer losses – around 300.

Strolling around the city, it was hard to imagine the brutality of the past. Despite the conflicts we have around the world today, it makes me thankful, at least, for less troubled times during my lifetime.

With various outdoor shops to hand, I secured new trekking poles and a battery pack. I resupplied with food and returned to the hostel, where I had booked in for another night to make the most of my rest day. The thankless task of removing packaging and transferring my supplies to plastic bags was the last of my jobs for the day.

I calculated I had around 90 miles left on the Cape

Wrath Trail until the end in Fort William. There I hoped to regroup, dry out, do laundry, and then resupply for the 100-mile stretch on the West Highland Way.

I admit to being fed up with the CWT at that point. The poor weather (it had rained at some point every day so far) and rough terrain were taking their toll. But I needed to finish. Considering a short day due to the trip back to Strathcarron, I planned on completing the rest of the trail in four days. Any longer than that, and I feared I would lose my sanity. Assuming I could pull in a respectable distance from Strathcarron of at least twenty miles, that would leave me three days of twenty-three miles each to finish. Twenty-three miles in a day is usually easy for me, but on the CWT it was a completely different proposition. I had to get my head down, concentrate, persevere, and try to control whatever the hell was happening in my head. Having hiked seriously for the last six years, and extolled the benefits of the outdoors, it felt like unknown territory to be out in the thick of the wilds, doing something I loved, but yearning to be finished with it. Despite the confusion, I knew I had to get the hell off the CWT.

I'm usually a laid-back guy. However, I have a list of annoyances, or pet hates. Most may seem trivial, but we all have certain petty aspects of life that bother us. Allow me to offer a few examples.

The first is well-educated, intelligent individuals who, when common sense is offered, decline the invitation. They'll breeze through an explanation of nuclear physics, but ask them to check a bus timetable to find out what time number 37 departs, and they'll be stumped.

Second, anyone who uses phrases such as 'reaching out', 'blue-sky thinking' or 'it's on my radar'. I worked in offices for years – a small, unfortunate oversight in my life planning. Corporate phrases are all the rage, thrown into board meetings and office conversation, often by individuals hoping to impress and unable to communicate in plain English.

Replacing a perfectly adequate roundabout with traffic lights annoys me as well. I like roundabouts; they generally flow freely, and most drivers obey the right of way. Then, out of nowhere the ominous roadworks materialise, and two weeks later there are more red lights than Amsterdam on a Saturday night. Joining the queue that never used to be there, we wait patiently for five minutes for the red to change to green, balancing the throttle, handbrake on and first gear engaged for a quick getaway. At worst the first driver in the tailback is distracted by a tune on the radio and fails to move. At best, maybe four cars make it through before the light changes back to red. You can spot fumes not only from the exhaust pipe but from drivers' ears as well.

Fourth on the list is the beeping sound made by car reversing sensors. It's way too impatient and needs to relax. My current car is the first I've owned with this extra, and, to further aggravate the situation, I can't turn it off. Engage reverse and it likes to confirm you have done this with one, single beep. Inch back to the vehicle behind and it speeds up, quickening until reaching the point where it suffers a seizure. It's worse than a passenger directing you back. "Yeah, you're good. Slowly, slowly, easy, I said EASY. Watch it! Be careful! Are you nuts?! No! STOP! For the love of God STOP!"

Next, instant coffee. Where do I even start? Becoming a true coffee connoisseur takes years. I'm still on the journey, but every year I make a little tweak on my path of bean enlightenment. This started years ago with instant granules, for which I lay the blame firmly on my father. The only positive side of instant coffee is that things can't get any worse.

During my first trip to the States in the mid-nineties I felt I'd landed in bean nirvana. Every petrol station had rows of piping-hot coffee in flasks. OK, so it was a flask, but compared to what Esso was offering back home (a vending machine dispensing a sealed cup of Gold Blend) it was a revelation. I cottoned on to the fact I could not only refuel my car, but my body also. Cafés were springing up everywhere in the bigger towns. My first visit to Portland, Oregon, renowned for its abundance of cafés, had me skipping around like a hyperactive caffeine junkie for hours.

The UK took ages to catch up, and it's only been in the last 10 years that our caffeine availability has improved. But, coffee granules still lurk. I know where my good local caffeine sources are, but I get caught out on trips and venture into unknown establishments. Requesting a black Americano (half full with an extra shot) is met with confusion and the terrible reply, "We have instant?"

Here's one only the British will relate to: traditional glass milk bottles with a foil top. I may draw blank looks from my American readers, but we can still get our milk delivered in the morning from the local milkman. It's a great, traditional service, but the service isn't the problem. I believe everyone accepts the humble pint bottle because it dates back years, and no-one questions it. However, the design is flawed. A foil top requires a gentle push to remove it, rendering it useless because it either falls into the bottle, drops off when you even dare look at it, or fails to offer any form of seal once opened. However, it's the bottle that annoys me. The milk is filled to the brim, so a careful pour is required, especially the first one. I don't drink milk any more except in a cup of tea so only need a splash, and the tilt must be gentle so as not to dispense too much. Try as I might for a gentle pour, the contents merely dribble pathetically, half in my tea and the rest down the side of the bottle, over the cup and onto the kitchen counter. Milk bottles – they're out to get me.

Manicured poodles are my penultimate pet hate (excuse the pun). While a poodle, in appearance at least, wouldn't be my first choice of pet, this poor canine suffers at the hands of those owners intent on just plain embarrassing it further in public. Can I say to all those pruned poodle owners out there that shaving your dog's legs, neck, arse or wherever does nothing for the appearance. You'll have upper legs shaved down to skin, with two white footballs just above the feet, or a tail that looks like a strip of wire with a white Russian Cossack hat bouncing around on the end.

Finally, in pole position, is paying good money for poor meals. I used to work as a cook, and, as in any service industry, the premise is simple: a customer pays you money for a product which naturally we expect to be worth our cash. It isn't a difficult concept to grasp, but some establishments still don't understand it.

Case in point just outside of Strathcarron. I disembarked at the train station at 10am, not too late to pull in some respectable mileage, but I hadn't had the chance to eat anything after rushing to Inverness station. The trail followed the A890 for three miles, and, not being the quietest of roads, it proved a game of dodge and dive in the hedge every few minutes to avoid traffic.

Before long the Carron Restaurant appeared to my left. It wasn't busy that morning, so I hoped to grab a bite to eat and be on my way. Four others were dining as the cook tended to a couple of steaks on a flaming grill, the smell of which determined my choice of brunch. I opted for a rare rump, sliced and served on a bed of 'seasonal leaves'.

Now I know when I cook a steak at home, depending on the thickness, that two minutes each side will give me some pink in the middle, just how I like it. A further five minutes to rest the meat and throw some salad leaves on the plate, drizzle a little dressing and make my Americano (half full with an extra shot) suggested a timespan of around ten minutes. As he was already cooking for diners who had ordered ahead of me, I calculated 15 minutes would suffice. When my dish arrived five minutes later as I sipped my coffee, my doubts surfaced.

I checked the plate. Lurking under the fresh rocket was the out-of-date stuff, wilting and sweating as if nervous I had spotted it. The steak was overcooked; cold and re-solidified fat along one edge wasn't doing it any favours either. I prodded, unconvinced, and looked back at the waitress, who had retreated to the safety of the fridge freezer, looking back occasionally, one eyeball peering to estimate her chances of getting away with it. Another diner looked in my direction with what appeared to be an expression of alarm, or at least warning.

The meat was stone-cold and tough – just one slice took an eternity to work my way through. As I chewed it became clear that my hiking schedule may suffer an unexpected setback. The further I progressed down through the pile of rocket, the less appetising the leaves became, turning from a crisp, vibrant green on the surface to withered slime. I caught the waitress glancing as I gingerly lifted each past-its-best leaf to drape it over the top of the plate, as if offering visual verification of my displeasure. Somehow, I ate the meat but left the rest.

"How was your meal?" she asked, returning to collect my plate.

"Er, to be honest?" I offered.

She didn't reply.

"Not good," I continued. "The meat was cold and dry. Most of the rocket is bad."

"I'll make sure the chef is made aware," she finished, and returned to the kitchen.

Despite expecting an apology or a concession on the bill, neither was forthcoming, so I left. I've since heard that the Carron was taken over after my visit and now enjoys rave reviews.

The sweet smell of seawater drifted over from Loch Carron as I continued down the road to a turn-off by Attadale Gardens, where a narrow side road continued two miles to the River Attadale. At the request of two other day hikers, I took photos for them, and we talked briefly about our respective walks.

After reaching the river, which signalled the end of the road and my return to wild Scotland, I made good progress on a gently climbing track next to a forest. Just shy of Loch an Droighinn the trail spun right, merging into the wonderful Glen Ling.

Tarmac greeted me before the quaint little hamlet of Killilan. I was making good progress; with the road walk out of Strathcarron, the good state of the trail in between, and now being back on firm ground once more, my mileage was increasing for the day. I rested on a bench by a phone box in Killilan and ate some snacks then lay on the grass soaking up the sunshine and looking forward to more road walking.

Most hikers either love or loathe road walking. I love it but only for short durations, say up to five miles. The main positive side to grinding the tarmac is the freedom to look around. When I hike on trail, I focus my gaze mostly on the ground six feet in front, which I need to do to concentrate on my foot placement. I invest a lot of time and money in thru-hiking and don't want my adventures to end because of carelessness. During my 12,000-plus miles of hiking, this approach has served me well; I've fallen just once and remain free from serious injury.

But I often find I've walked for a few miles and remember little of my surroundings. My recollections of the forest, the desert or wherever I've just travelled through are weak because I was focused on the trail. Over the course of the past few years, I have experimented with where I look and how I do it. The method I now use to balance between paying attention to foot placement and experiencing my surroundings is to home in on my sweet spot. I hold my head upright to avoid neck strain when looking down and use what I can best describe as a 'vague' gaze. There's no direct focus on anything, but my concentration aims forwards up the trail. My peripheral vision plays with the ground ahead, and, if any trail detritus warrants further attention, I look down and check to see if I need to adjust my stride before returning to this central gaze. This method also provides the chance to experience my surroundings more, and I've discovered I can see more of the countryside I'm hiking through and, more importantly, remember it.

As crazy as it sounds, I've honed this approach further with a little yoga. Amongst the usual physical exercises you might associate with yoga, there are smaller and more precise adjustments. One of these is the eyes and the ability to develop this field of vision. Many yoga poses concentrate on the spinal area, right up to the neck and the head. When the head is stretched backwards, forwards, or side to side, the final part of the position is to also turn the eyeballs as far as possible in the same direction. This works the eye muscles, and I've found my outermost field of vision is far better for it. The net result of all this work is that I can merrily skip along the trail with a decent balance of paying attention to the surface and remembering my surroundings.

On a road walk I can truly relax, because I don't use my field of vision so much. The surface is smooth, I'm not going to trip over anything except the odd pothole, and this enables me to look around. I relax on road walks, fall into daydreams, and miss turn-offs or trailheads as a result.

Unfortunately, there's a negative side to the freedom. Road surfaces are hard. Off-road the trail is kinder; it's softer and absorbs some of the ground shock, which makes injury less likely. Plodding along on asphalt for short periods is fine, but do it for hours and there can be repercussions.

Because it's difficult to plot a course over private land or unnavigable terrain, the Continental Divide Trail utilises long sections of road walking. A lot of the long-distance trails in the US have road sections, but, during my brief time on the CDT, I experienced lengthy spells, sometimes up to a day. The initial novelty is welcome – I can relax, look around me and pull immature faces at car drivers – but the novelty soon wears thin. Legs, back and feet take an absolute pounding. I've covered distances of up to 1,000 miles without a blister. A day on the road, and I'll reach camp in the evening sporting two or three new ones. My legs ache, my back hurts, and all I want to do is lie down for 30 minutes waiting for the pain to subside.

I also get carried away and increase my speed, knowing I can crack out impressive distances. My usual pace on a good trail over level ground is around 3.3 miles per hour. On a road section I can take this up to 4 or 4.5. I revel in the new-found freedom, ignore the possible side effects, and let rip. The progress gained further fuels this state of mind. When I check my distance and see that 8.5 miles have passed in two hours, the endorphins party, only fuelling the addiction.

One afternoon section on the Appalachian Trail, I was trying to catch up with some hiking buddies and had 18 miles to cover to reach the Rod Hollow shelter for the night. I'd swallowed a couple of ibuprofen to ease up the muscles and delay the inevitable hurt, plus a mouthful of potent energy drink that was trickling down to my leg muscles. With little wind, relatively flat terrain, and the trail in good repair, I was flying. I pulled in to the shelter having covered 18 miles in 4 hours. The fallout, confirmed by a running coach at the shelter, was a shin splint. Shin splints are exercise-induced pain to the shin area, and severe cases can require days of rest. With more ibuprofen, minimal movement that evening and several visits to a nearby cold spring to soothe the injury, I was fine by the morning. But I learned my lesson.

Getting older doesn't help either. I need to put in far more work over the winter months before summer to keep the fitness I've gained. I can still pull in distances of over 40 miles a day even now, but I must be careful. I need full days off to allow my muscles to recover when training, and I must pay attention to my pace and warm up slowly. When I am hiking fit and fully loaded, 25-mile days are easy, 30 presents no problems, and it's only when I hit 35 miles or more that my body starts to complain.

The CWT presented no injury risk on road sections because they weren't of significant duration. It was a perfect balance; I could put my foot down to satisfy my endorphins and be back on trail before any damage was done.

I'll be a happy man if I'm still pulling in those distances in another 20 years.

Chapter 7

Reflections in the Cluanie

Depression is an illness, a mental illness – we need to get that out of the way first. It's a common misconception that those who suffer with depression will 'get over it', 'snap out of it', or be fine the following day. As with many medical conditions, often there is little we can do about them. Some problems can be remedied, such as taking an aspirin for a headache or antibiotics for infections. In between these two extremes are situations where we can't always cure the problem, but can take positive steps to ease the symptoms and be able to live a relatively normal life. There is no cure for depression, but it can be treated. Lifestyle changes such as healthy eating, plenty of exercise, and exposure to sunlight can, in many cases, help enormously. Some choose medication such as antidepressants, although the side effects can present more problems. The single most important step many experts consider vital is to see a counsellor.

Depression is widespread. Statistics vary, but it is

commonly accepted that one in six of us on average will experience depression at some point in our lives. Sufferers experience low moods, feelings of guilt, poor self-esteem, lack of interest and enthusiasm, and little hope for the future. It affects sleep patterns, often preventing a good night's rest. Many sleep during the day to escape the fears that manifest themselves when awake. The ability to concentrate on simple tasks such as washing up or even getting out of bed deteriorates. It influences the appetite – sufferers often have little desire to eat at all, but at other times they consume nutritionally poor feel-good foods. There's a general mood of helplessness and lack of faith in the future, with no hope for tomorrow, the next few days, or even years. In worst-case scenarios, sadly, those who cannot deal with depression take their own lives.

The causes are still not fully understood, but we know there is no one single reason. Often circumstances such as a bereavement, redundancy, illness, or other events can trigger it. More often than not a combination of various factors compound and develop into depression. Past emotional, sexual and physical abuse can also spark depression in later life.

To confuse things further, there are several forms. Mild depression may have a limited impact on your daily life, such as the inability to concentrate at work. Major depression is far more serious and affects simple, everyday activities such as sleeping and eating.

Bipolar disorder, also known as manic depressive illness, forces wild mood swings, from extreme highs where the individual feels elated with no problems, to the other end of the scale: complete despair, even suicide. Bipolar sufferers often display illogical or out-of-character behaviour.

Many mothers experience anxiety and lack confidence

after giving birth. Postnatal depression can leave mothers feeling overwhelmed and unable to cope, experiencing negative feelings towards their child. Often this begins two to three weeks after birth.

Although painfully aware that I wasn't in good shape, I didn't realise I was suffering from depression until the following year, when my doctor diagnosed the condition. Being ignorant about the illness until that point and unwilling to talk to anyone for fear of what they might think, I suffered in silence. I thought that I was just having a bad spell and that tomorrow would be a better day (which it often was). But my realisation, and fear, was that the pit, always ready to catch a wayward foot, lay just around the next corner.

I didn't just struggle on the actual days of depression. Often, on good days, it was the anxiety of not knowing when I'd sink back down again. The beast always lurked, constantly stalking me. I knew this monster was hiding in the grass, biding its time and enjoying the hunt.

The overriding fear was knowing it was going to pounce once more but not knowing when.

Leaving Killilan, I walked for two more miles on a small side road before it disintegrated into a rough track and funnelled me into Glen Elchaig. Down a steep bank to my right, the river Elchaig tumbled over rocks and cut a sparkling swathe through lush grass. Towering pines clung to the slope and rose high above me, swaying in a light breeze which carried cries from a flock of sheep.

Ignoring a turn towards the Falls of Glomach, as tempting as the route sounded, I carried on past Loch na

Leitreach. The far shores rose grandly to the summit of Meall Sguman. The small hamlet of Carnach, comprising just a few livestock buildings and an abandoned farmhouse, signalled the turn I needed.

The CWT is not a one-trail route. It splits into variants at intervals along its course. Checking the map over the course of a day, or planning the route in the evening, I sized up each choice. I was still picking the shortest route possible, battling my emotions to finish the trail and get onto the firmer footing of the West Highland Way. This wasn't as easy as it sounded.

The shortest route was not necessarily the quickest, and often the longest course wasn't the wisest. Faced with the choice of a path of limitless Guinness or dystopia, I tended to veer towards the familiar solid red dashes. I preferred more mileage on a decent surface than less distance tramping through boggy ground. For example, the turn I had ignored towards the Falls of Glomach promised a longer stretch of good trail, but it swung a wild loop out west to the village of Morvich then doubled back east again. My route from Carnach was shorter but appeared speckled with red dots and the dreaded black ones. It was never simple – I always faced a trade-off somewhere.

Fort William lay just 51 miles ahead of me, and from Invergarry, another 47 miles distant, I was assured of a good trail. From there, the CWT met up with Loch Oich and followed the splendour of the Caledonian Canal, past the vast Loch Lochy and all the way to the finish at Fort William via the Great Glen Way. The CWT shared the Great Glen Way, which, being more popular, sported a well-maintained and flat trail beside the canal. Joy fluttered in the pit of my stomach when another glance at the map revealed several little blue cup symbols dotted along the glen, which meant one thing – coffee.

A mere 15 miles away was Loch Cluanie, where the trail showed a marked improvement in condition, and which was home to the Cluanie Inn. Tomorrow I could be enjoying a late breakfast at the inn and contemplating just a couple more days before I had conquered the most difficult trail in the UK. My mood lifted, and my depression sank back into the undergrowth.

As I set up camp by the river near Carnach, the elements threw in a pre-finish surprise. Not only did I have a great spot for the night next to the Allt Coire Easaich at the head of Loch na Leitreach, but Scotland's weather pattern served up a spectacle that matched my mood: promising.

It developed into one of those evenings at the end of a long day when everything about the great outdoors falls gloriously into place. At a rough guess, I've spent a total of two years of my life hiking and camping. I'm not able to remember every single moment. Some memories have left me, others return with a little concentration. A few, however, shine.

That evening was one that glistened. Even now, two years later, I can picture it vividly. From the slight slope of my camp, the slippery bank to the river I negotiated for water, rocks peeking through the moist grass, a darker shade of their usual grey after a passing storm, everything was perfect.

The midges seemed occupied with other tasks and left me alone, and sheep eyed me curiously, more focused on finding the tastiest grass. I sat on the banks of the Allt Coire Easaich, my feet in the water, soothed by the chill. Looking west down Glen Elchaig along Loch na Leitreach, the glen narrowed towards the mighty obstacle of Carnan Cruithneachd before disappearing.

An early evening light softened Scotland. As the sun

edged closer to the mountains, the sky gradually changed from a northern blue to a theatrical orange. Maturing before my eyes, vivid scarlets joined the party streaking across the horizon, and the blackness of night glanced the upper reaches. The slight afternoon breeze conceded and surrendered, as though resting to start again fresh in the morning. These colours, an artist's palette, reflected in Loch na Leitreach. As if the scene above the horizon weren't enough to satisfy me, the still, mirrored waters below offered even more.

Aware of my sporadic unhappiness, in moments such as that evening I forgot my problems. Even on the difficult days, Scotland sensed my sadness and occasionally threw light my way, an outstretched arm of support. Perhaps it was suggesting, whispering I needed to accept I had problems. If I could accept that admission, then I could begin to resolve it.

Rain fell intermittently overnight but stopped just as I decided to get up. Fingers of an early morning mist clawed and fumbled up high – a giant's hand grasping the summit ready to haul itself up. Droplets of dew speckled the grass, a million eyes glistening and studying me. Already the sun soared above, beginning its long arch west, drying out Scotland, which steamed in gratification. Despite being the middle of July, snow still lingered on the north faces as low cloud brushed over them.

I set off in my T-shirt – bracing for the early morning chill – and eyed up the steep climb, which I hoped would warm me. The trail disintegrated into tussocks and steepened. I struggled up the incline while searching for an easy route over the uneven ground.

Topping out I picked up the line of the Allt Coire Easaich once more, which I knew led to Loch Lon Mhurchaidh. The river crashed and bubbled, swerving and curving like me as we both eyed up the terrain and adjusted our courses to pick the line of least resistance. I passed the loch and, checking the map, aimed for the Abhainn Gaorsaic River, which threaded an easy course through a gloriously wide, green valley. Lochs Thuill Easaich, Gaorsaic and Bhealaich all nodded confirmation of my route. The peaks either side, A' Ghlas-bheinn and Sgurr Gaorsaic, appeared gentle; lush grass lifted from the lochs and clung to their flanks up to the summits. My valley seemed welcoming as it steered me eastward into the majestic Gleann Gniomhaidh.

Working my way down the glen, I spotted a small building in the distance, corresponding to a familiar youth hostel symbol on the map. I strained to see it from three miles away, but as I progressed the corrugated green walls and brown roof of the Glen Affric Hostel became clear. A wisp of smoke rose skyward from the chimney unhindered.

Strolling over flat, easy ground, the surface crunching under my shoes, I stopped at the front door and set my pack down. A sign taped to the inside of the window said the hostel was open and, underneath, another note made me smile.

Tea and coffee available against a small donation.

I removed my shoes and ventured inside.

"Hello?" I offered in the quietness.

"Hello! Come in."

I made out her outline in the gloom. She waited for me to approach down the hall.

"Hi, my name is Fozzie. Sorry to disturb you, but I couldn't pass by without taking up your offer of a cup of tea."

She laughed.

"I'm Audrey, Audrey Leonard. Of course, come into the kitchen."

She turned on the kettle, and two cups clinked as she placed them on the counter. I looked around the kitchen. It was cosy; sunlight streamed through the window, and I sensed the warmth on my back as a black iron stove smouldered in the fireplace, surrounded by smoke-blackened rock. Wood cladding covered the walls, reaching up to a faded white ceiling. Towels hung from a line above the stove, and a red, chipped fire bucket sat on the floor. I felt I'd travelled back in time 50 years.

She took the time to sit with me, and we chatted about my walk and her job looking after the place. She enjoyed the remote location and looking after the collection of outdoor folk who regularly stopped by for refreshment, or to stay overnight.

Warning me of bad weather and possible storms that evening, she pointed to the trail up the valley towards Loch Cluanie.

"Go over the bridge and keep to the fence line for as long as you can," she advised. "It's a rough track, but you'll reach the loch this afternoon. Maybe before the storm hits, you can shelter in the Cluanie Inn."

I left with a fine cup of tea fuelling me and thudded over the bridge spanning the River Affric to start the climb. The day had lifted me. A series of small details made me smile; the perfect camp that morning, occasional sunlight breaking through, the inspiring Glen Gniomhaidh, and finally a cup of tea with a welcoming host.

Pausing after the bridge and wondering if the day could get any better, I stood in awe, taking in the landscape around me – a magnificent focal point, a convergence of four spectacular Scottish valleys. Captivated, I didn't know which direction to look, each glen offering its own perspective. At times my surroundings seemed surreal. Four rivers, the Allt Gleann, Allt a' Bhuic, Allt a' Chomhlain, and Affric all snaked through each valley, sections glistening in the distance as intermittent spotlights of sunshine caught them. The might of Ciste Dhubh rose, directing me onwards, and after a few minutes I headed away. Like a postcard offered as a memory, a parting gift, I knew I'd return one day to experience that place again.

I climbed into An Caorann Mor, remembering the warden's advice to keep to the fence line. The ground fell and rose, submerged in small pools of water. Breathing heavily, I resisted the urge to look ahead towards the col, as it never seemed to get closer. I glanced back, the bridge and hostel shrinking each time. Intermittent sunshine teased the temperature, and I struggled to find a comfortable balance of clothing, overheating as sweat ran over my face.

The white exterior of the Cluanie Inn tempted me over when, seven miles later, I reached the road. It was early evening; as I looked south towards the next part of my route, I knew the CWT wouldn't let me go without a fight. The track skirted Loch Cluanie for a while before climbing and disappearing into low, threatening cloud. I remembered the storm warning, counting myself lucky that at least my day's walk was over and I'd be able to sit out the worst of it overnight.

Cars hissed on the wet road as I walked to the inn. I scanned the menu on the window and peered inside to see if they were busy. A few people sat at tables drinking and eating; I debated whether to enter. The rain started and made my mind up for me.

Splashing water on my face in the Gents, I looked at my

reflection. Two hundred miles had taken their toll; my hair was matted, dirt streaked one cheek, and my face was sunburnt. I looked tired, drained, but it wasn't a physical weariness. My eyes appeared weak and unsure. I'd never seen them like that before; there was no sparkle, merely a resigned, faded gaze. I stared into the mirror for a while, motionless. Who was this person scrutinising me? He looked lost, confused, and sad. For a moment he seemed to plead, crying for answers to questions he didn't yet have. He wanted help.

I wiped the mud off my face, ran my hands through my hair and rubbed soap through my beard before going to the bar. After ordering food and a beer, I sat in the corner looking at nothing, focused on my meal and drink. The waitress returned to collect my plate a while later.

"Everything OK? How was your meal?" she asked.

I couldn't find the effort to reply.

"Sir? Was everything OK?" she repeated.

"Sorry, yes, it was good, thank you."

"Can I get you anything else, another pint?"

"Yes please."

I left at closing time, once more several beers and whiskies the worse for wear. My reflection in the mirror on my mind, the memory of another person I didn't recognise constantly looked at me. The drink hadn't pushed the image away as I'd hoped; it merely haunted me further.

The rain continued to fall as I searched for a camp spot. I trudged the mile back to where I had arrived at the road, remembering a derelict barn I'd passed earlier. Standing inside I looked up to see rain streaming in through gaping holes in the roof, soaking the ground. There was little hope of staying dry, and the floor was concrete, so I couldn't pitch my tent either. Returning to the inn, I carried on towards a bridge over the River Cluanie. A level area to one side offered the only flat spot, and I set up my tent, oblivious to the occasional walker passing me, not caring whether they thought I should be more discreet.

The clouds had sunk further, obscuring half of the mountains ahead, and a fierce wind ripped through the valley over the loch, turning its waters wild. I retreated inside my tent to peel off damp clothes, inflate my mat, and pull my sleeping bag over me. But I was not inclined to arrange my gear and supplies, which normally I was particular about.

I remembered most of the night because I didn't sleep. The rain didn't ease, pattering on the tent walls as a wind gust tore at the guy lines. I yearned for a cigarette, and, as the next day eventually arrived, I felt no motivation to get up and hike. I had just one more day to reach Invergarry where the trail improved, but my mood pinned me down, refusing to let me go. The storm had passed, but a low, grey sky covered the outside world, matching my frame of mind.

Reaching by my head I turned the valve to deflate my sleeping mat, leaving the ground cold and uncomfortable, forcing me to get up. I packed lethargically, every movement and action slow, requiring all my concentration and effort. After packing my wet shelter away, I took one look at the trail disappearing up into the clouds and walked up to meet them.

I nearly turned around. Maybe I'd rather return to the inn, eat breakfast, have some coffee and call a taxi to take me somewhere with a train station. I'd had enough. My head thumped from the beer as I rubbed my forehead in a vain attempt to ease the pain. I fumbled in a side pocket for the ibuprofen, but they'd run out.

I climbed, lifting my head from its bowed position to check on the advancing cloud base. I wanted to enter the grey, for it to envelop me so I'd be forced to complete the last difficult section just to get out the other side. Continuing up I pulled on my poncho and left the stony road to trace a passage around the lower flank of Creag a' Mhaim and reach the River Loyne, the halfway point to a small country road into Invergarry, where the trail improved. I passed another hiker coming the opposite direction, but neither of us stopped, just offering small acknowledgements and wet, weak smiles. I wondered why he looked so sad and pondered if he thought the same of me.

The route passed north of Creag Liathtais and swung out west to meet the Loyne. I paused, looking at the river, and contemplated scrambling down rough ground alive with cascading water to avoid the detour, but I relented – it looked far too dangerous.

I had to cross the wide river, but thankfully, even with the rain, it remained shallow. The trail, doubling back on itself from the detour, followed the bank for a mile before the crossing. Without breaking stride I waded in, the water reaching my knees as I fought the current to the other side. Unable to find low ground, I clambered out on my chest, grasping anything capable of pulling me out.

Again, I ventured higher and aimed for the pass at Mam na Seilg, also obscured by clouds. The path comprised little more than occasional rock and soaked peat. Water swept over my shoes, and I couldn't even muster the concentration to check the map, instead following the approximate direction to the col, hoping my bearing was correct. Loch Loyne stretched out to the east, occasional sunlight dappling the far shores, giving me hope for a dry end to the day.

I clung precariously to huge, slippery stone slabs as diverted burns raced over the surface. At times I didn't dare move, each shoe teetering on a tiny point of grip that I feared wouldn't hold. My fingers scrambled for any morsel of rock capable of offering security, and, when found, I tentatively moved a foot to the next placement until I reached the mud once more.

Finally I topped out at Mam na Seilg and looked down onto the Glenquoich Forest. I could see the river flowing into Loch Garry. The small bridge I needed to aim for was, I hoped, the point where my trail improved into Invergarry.

I was nearly there. The Cape Wrath Trail was almost finished.

Chapter 8

Elina

few years ago, during a period in my life which I refer to as 'sporadic travelling', I worked various temporary jobs in the UK, earning enough money to travel to other countries. What I was escaping from I didn't know, but I knew what I was searching for: contentment.

Working a series of dead-end, mundane positions, I cleaned cars, toiled in kitchens, filed documents at the local council, and collected people's refuse. I hated those jobs – they were meaningless and often just compounded the initial problem of unhappiness – but they provided enough money so I could travel. Those destinations were places I could relax with no schedules and spend a few months figuring out where my life was heading and what I needed to do to find peace of mind.

I felt free during those escapes. I experienced wonderful countries in Europe, visited America and Canada, and met amazing people on their journeys. But still something was missing; despite my searching, contentment or clues to discover it were not forthcoming.

Sitting in a bar one evening in the city of Heraklion, on the north coast of Crete, I got chatting to a Kiwi called Bob. I told him that I wished to spend the last of the summer walking but was unsure where to head. He mentioned a friend who had just returned from hiking a route known as the Camino de Santiago in Spain. Curiosity put up a decent argument for getting the better of me, and I allowed it.

A week later I stepped off a train at Le Puy en Velay in southern France, tightened my laces, and began walking 1,000 miles on El Camino, west to Santiago de Compostela. By the time I had finished, many life questions were answered, including the most important one. That lesson was simple: the destination holds little significance, it's the journey that matters.

I returned from Spain a different man. Convinced that I had always been heading to one defining moment, despite protestations from friends that life just didn't work like that, everything became clear. The Camino taught me to enjoy life, to experience my odyssey, and to not be afraid of altering my path – no matter how risky those changes seemed – if I was sure they were for the better. It was then I decided to spend more time hiking and to write about my adventures. These two, simple goals changed me; I was content.

But now my contentment was running dry. Like a mountain spring that, over the course of a summer, weakens to a mere trickle before vanishing, my well-being had diminished from a sparkling torrent to nothing but dust. If I could have scooped a handful and let it slip between my fingers, I would have.

During my hike on the Cape Wrath Trail, and my other hikes, I'd had plenty of time to think – often a blessing, sometimes a curse. There'd been a confusing mix of high and low, and I realised my emotions had been running up and down like an unstable seesaw for a while. Since five years ago to be exact. Ever since my thru-hike of the Pacific Crest Trail, which, at times, had reduced me to tears, I had kept returning to this dark place.

I was on a rail journey where many stations called *Contentment* flashed past in a blur, but the train never stopped. Eventually, the carriage lurched harshly and came to a standstill. I looked at the station sign.

Dark Place – End of the Line.

As if the name weren't foreboding enough, the conductor added to my woes. "Last stop. We terminate here, everyone off please."

I didn't know what was wrong but couldn't ignore my feelings. I had to do something about it.

I reached the River Garry and sat, breathing hard, sucking in oxygen until my chest relaxed and heart rate returned to normal. I hadn't finished the CWT, but I knew the difficult part was over. No more bog, no more constant wet feet, and I hoped that my mood might improve.

I crossed the bridge and weaved along forest fire track for a few miles with the Garry occasionally veering over to say hello. It was mid-afternoon, and the sun was high, bathing the trail in light before the pines blocked it again. I sensed the temperature changing, increasing my speed through the shadows to reach warmth again, thankful the rain had stopped.

I had 13 miles to Invergarry, where I hoped to eat at the hotel before finding somewhere to camp. After that it was a mere 32 miles to Fort William, the end of the CWT and the start of the West Highland Way.

I sped through the forest, lifted by thoughts of another trail under my belt and a decent meal at the hotel. My world, surrounded by trees, blocked out the familiar views I had become used to of mountains, valleys, and lochs. My feet crunched on smooth gravel, my focus lifting as I didn't need to concentrate on foot placement.

I felt out of place when I arrived at the Invergarry Hotel's smart and clean interior, over-conscious of my appearance. I went to the toilets for a quick repair job and wetted my hair as it retaliated wildly. My hands were grimy, dirt collected under the fingernails, and Scottish bog streaked the inside of my ankles. My reflection in the mirror appeared as it had done at the Cluanie Inn; but this time I smiled at the guy looking back at me, and he reciprocated.

It was Friday evening. The bar was quieter than I had expected, and I ordered a Guinness while checking the menu.

"Can I have the fish and chips as well please?" I asked the barmaid.

"You can, my love," she replied, before adding: "You look hungry! Take a seat, I'll bring it over for you."

I lowered myself gently into a chair, tucked discreetly in a corner, but when I began to relax my legs stiffened, seizing. I let them, aware I had just one mile to find somewhere to pitch my tent. Diners eyed me like a wild animal. One of them prodded her friend as if to suggest they talk to me. I turned away, not keen on conversation, as the waitress placed my meal on the table.

"More Guinness, sir?"

"Yes please."

Refuelled and watered, I stood up, rubbing my legs in an

attempt to convince them they were capable of another 20 minutes, and left.

My easterly bearing, blocked by the vast Loch Oich, swung south to follow the banks through more forest. The terrain sloped, offering little opportunity to camp. I reached the road and crossed it as a light rain fell. A narrow strip of land near the loch gave me the chance to set up my tent, before diving inside, eager for rest after a 29-mile day.

I woke to the pitter-patter of showers on the tent, but as I packed they stopped. Walking just a mile, a sign advertising freshly made bacon rolls and coffee tempted me over to the intriguingly named Well of the Seven Heads Store.

After eating, an obelisk by the side of the road roused my curiosity. The gory monument of a hand wielding a dagger with seven heads in a circle underneath stood by the shore, a gruesome reminder of past times when Scotland was a lawless and bloody place. The clans back then held no respect for regulations, and the authorities often ignored them.

In September 1663, Ranald and Alexander MacDonald of Keppoch were murdered by their uncle, Alasdair MacDonald, and his accomplices. Accounts vary, but it's thought the killings happened due to either a land dispute or a quarrel at a party after the MacDonald brothers returned from schooling in France.

Two years later, Iain Lom, a kinsman of the MacDonald brothers, took his revenge. With a group of 50 men they travelled to the MacDonald home at Inverlair and killed all seven of the murderers, decapitating them afterwards. Lom collected their heads and set out to show them to MacDonnell of Glengarry, who had failed to bring the murderers to justice originally. Lom stopped en route to wash them in Loch Oich, where the current obelisk, erected in 1812, now stands. In the late 19th century, headless skeletons were unearthed at the grave in Inverlair, backing up the legend.

It provided another reminder of just how brutal the world used to be, and Scotland in particular. I'm grateful for living in kinder times, but I wonder how, in another 200 years, civilisation will judge us? With wars still raging, terrorism, killing, and violence, will our world today be viewed with equal disdain?

The Great Glen Way (GGW) stretches 73 miles from Fort William to Inverness, following a major natural fault line. Much of the route hugs the shores of three large lochs: Loch Lochy, Loch Oich and Loch Ness. In between these, the Caledonian Canal fills in the gaps. The GGW shares its path with the CWT, and as I walked it Scotland transformed from bog and mountains to a flat, smooth path that appeared so well kept that I expected to round a corner and find a caretaker firing up a vacuum cleaner This wasn't so much a trail of limitless Guinness, more a walking wonderland that had popped up, offering a helping hand to my beaten-up body and mind. I was ecstatic, fuelled by my bacon roll (and another to take away), buzzing on caffeine and knowing I had merely 32 miles to the finish point. My world was flat, the sun blazed, and I felt alive. Screw my low mood; the Cape Wrath Trail was almost done.

I arrived at the Laggan swing bridge separating Loch Oich from the Caledonian Canal just in time for it to open, allowing a boat through. I waited patiently along with two cyclists and a few cars, content to observe. Crossing over, I joined the south bank of the canal for a mile before reaching Laggan Locks. A few boats, tied to the shore, bobbed gently as the owners relaxed on board, reading the paper or enjoying a morning drink. As I walked over the bridge at Laggan Locks, once more the trail switched to the north side before the vastness of Loch Lochy stretched away before me.

Loch Lochy is over nine miles long, with an average width of half a mile and a depth of up to 230 feet. I walked along its length for an eternity, shaded by forests of pine, inhaling their sweet scent. Lochy sparkled through the forest as the sun played on its surface, bouncing light into the shadows that glittered gloriously against the ferns.

I reeled in the miles on the flat trail, losing myself in a meditation of movement, mesmerised and unhindered. Loch Lochy narrowed at Gairlochy, where yet another bridge took me back to the opposite bank. From this point the Caledonian Canal would be by my side all the way to Fort William.

The canal is a feat of construction. Seeing such examples of engineering always makes me wonder how we'd manage such achievements nowadays, let alone when the canal opened in 1822.

The idea was to offer safe passage for shipping through Scotland, so avoiding the dangerous seas around the north coast, including Cape Wrath itself. James Watt first surveyed the route in 1773, but it wasn't until 1803 that an act of parliament was passed authorising the project. They hired Thomas Telford to build the canal with the help of William Jessop. The work was expected to take seven years at a cost, funded by the Government, of £474,000. Both these estimates were exceeded by completion, which took 12 years in total and cost £910,000.

With the defeat of Napoleon at the battle of Waterloo in

1815, the threat to the British Navy diminished. However, shipping increased through the canal during World War One, between 1914 and 1918, as naval vessels utilised the route to avoid German ships prowling the north coast of Scotland.

The most impressive and famous section is Neptune's Staircase on the outskirts of Fort William. A series of eight locks, each 180 by 40 feet, transports boats up or down 64 feet, taking around 90 minutes.

Queen Victoria took a trip on the Caledonian in 1873, and, with the advent of the railways at Fort William, Fort Augustus, and Inverness, coupled with the dramatic scenery along the canal's length, tourism flourished. The canal is now a Scheduled Monument attracting over a million visitors every year.

Despite approaching 30 miles for the day, I felt as if I'd barely started as I arrived at Neptune's Staircase. The flat trail, great weather, and relaxing effect of time spent by the water brought a smile I had sorely missed. A few boats dotted the staircase, and tourists observed from the banks, cameras or ice creams in hand.

Working my way through the houses on the edge of Fort William, I hit the Narrows, a section of water that runs into Loch Linnhe, then into the Firth of Lorn before the ocean itself. Tiredness took hold as I navigated through the streets and roads, searching for the true finish of the CWT. I skirted the coast, wishing the end would arrive so I could find a place to eat and somewhere to sleep.

It was early evening when I arrived at the Loch Linnhe ferry. I set my pack down, rested my arms on the railing, and stared out to distant mountains. Seagulls cried above me, and a gentle onshore breeze tugged at the rigging of boats, their flags flapping.

Proud at having finished the toughest hike in the UK, I reflected on the last 16 days and 245 miles as I threaded back through the streets of Fort William searching for lodging. My head skipped around, flicking between emotions. Feeling positive with my success but sceptical of my ability to continue. Happy with my accomplishment, sad the Cape Wrath Trail was over. Confused with my state of mind, but eager to delve further into my psyche and find out why it was running riot. Keen to carry on to the West Highland Way, hesitant how I'd handle the next part of my journey.

As a light drizzle increased to pouring rain, I walked up the drive to the Backpackers Hostel on Alma Road, checked in, and, too tired to wash, collapsed on the bed and fell asleep.

I'd been sucked in to the Cape Wrath Trail, knocked around, and after just over two weeks, been spat out the other end looking somewhat dishevelled.

The West Highland Way (WHW) was Scotland's first long-distance trail and is also the busiest. Although it's prone to some wild weather patterns, and the landscape is rugged, it is more forgiving than some trails. Local authorities maintain the route admirably, keen no doubt to please the hordes of outdoor lovers from the UK and abroad who come to tick it off their hiking wish list. The condition of the path is excellent and it drains well, accommodation options are many, there're plenty of chances to eat and drink, and direction markings are numerous. The sheer numbers who visit every year indicate its reputation. While I wasn't looking forward to the crowds, I was eager to experience a kinder trail.

Despite having walked none of it, I'd often seen it snaking through the landscape on trips to Scotland, in particular on the drive through Glen Coe on the way to Fort William. Regarded as one of the best sections, I was eagerly anticipating Glen Coe from the perspective of a hiker, for once, and not a driver.

I woke to a room full of German school kids, all 12 of whom had come into the dormitory the previous evening, showered, dressed, gone to eat, and returned without waking me. They had just completed the WHW and relayed their adventure to me – euphoric at their achievement – between my trips to the laundry to wash and dry my sodden clothes. The news was as expected: glorious scenery, plenty of places to find food and rest, and even the weather appeared to be improving.

I scrounged two eggs from the staff, scraped mould off a slice of bread, despite warnings of edibility from the French woman who gave it to me, and helped myself to a complimentary cup of tea from the hostel. Pleased with my freeloading, I smiled and thanked a Dutch cyclist who donated a rasher of fried bacon to the cause.

Fort William is an outdoor Mecca. Apart from the thousands rolling in off the WHW and other trails every year, Ben Nevis, the highest mountain in the UK, is also nearby. The town acts as a base for many hikers and bikers in the summer, and during the colder months it is one of the best places in the country for outdoor winter pursuits. The town offers everything to everyone: countless bars and restaurants, while accommodation and gear stores cater to every whim.

Fed, caffeinated, washed and showered, I walked the short distance back to Fort William High Street and resupplied with some dinners and snacks for the next few days. I relaxed for two hours, reading the newspaper over a coffee. After calculating my calorie expenditure, I figured another cooked breakfast was thoroughly deserved. My plan was simple; make the most of the WHW, slow the pace, and take a leisurely five days to complete its 95 miles.

I spent 30 minutes finding the start, which I'd been convinced was on Belford Road near the River Nevis before the Tourist Information office pointed out it had been moved a couple of years before to the other end of the High Street. Being the stickler for detail I am, I tramped all the way back, then turned again to retrace my steps.

The new point sported a bronze statue of a seated man, with benches and lighting. I found out months later, confirmed by many sources, it was moved so hikers would need to walk through the High Street. Before this change, the route finished before the shops, so the decision was made to improve business in the main street!

The bustle of town faded as I crossed the bridge over the River Nevis and carried on down a quieter road through Glen Nevis. Occasional gaps in the trees revealed Ben Nevis itself, soaring up before disappearing into the cloud base. I picked out the track I had climbed six times before during previous ascents. This time, for once, I wasn't here to scale the mountain – a less energetic day lay ahead.

Turning off the road, once more my route returned to countryside. I rose gently into the Nevis Forest, my feet cushioned by pine needles and a soft path. The surrounding hills, carpeted in greens, rose to jagged rock faces. The sound of cascading water grew louder as I passed a waterfall; looking up I watched it tear down the rock before slowing as the incline tempered, disappearing through drainage channels under me. As the gushing faded from one, the next intensified as I approached, and so the pattern repeated.

My narrow path widened to a broad, stone track, and I broke through the treeline, peering back down on the glen. A stationary cloud bank obscured everything above me, a colder world of rugged peaks. Light rain fell but struggled to make it into the forest. I put up my umbrella to deal with the persistent drops that made it past the canopy. So far the hordes of hikers hadn't materialised, and the WHW was failing to live up to its crowded reputation. Despite an occasional section of mud where the trail dipped and collected drainage, my feet remained dry.

I passed a turn-off to Dun Deardail, an Iron Age hill fort dating back 2,000 years. In keeping with my decision to pursue a more relaxed hike, I went to take a look. Reaching the summit of Sgorr Chalum, a solitary, grassy knoll, I sat and looked around. The views were expansive in every direction, and it wasn't hard to see why the fortification was built in such a place, taking advantage of the height and dominance it offered over any hostile forces.

Archaeological excavations have revealed a rock wall circling the site. Some of this stone is vitrified, suggesting the fort was burnt deliberately. Vitrification, or the melting of rock, only occurs with sustained heat, so experts believe that, not only was the garrison set on fire, but the blaze was stoked and kept alight to destroy it.

As I contemplated carrying on, a woman emerged from the forest below and approached.

"Hi! Good morning!" she said, breathing hard and rosy cheeked as she came and sat right next to me. "My name is Elina."

"Hi, I'm Fozzie," I replied, holding out a hand to shake.

I guessed she was around 35. Blonde hair, slightly curled, fell around her face and rested on her shoulders. Her skin was pale, dotted with a few freckles on her nose. She removed her red jacket to cool down and used it to sit on. She took out some oatcakes and munched on one, offering me another, which I accepted.

"What do you do here?" she asked.

"Just resting," I replied. "I'm walking the West Highland Way to Milngavie. You?"

"I leave Fort William for a short section on this trail as well. I go to Kinlochleven today, then back to Fort William to stay tonight. Tomorrow I return to Kinlochleven and continue."

She spoke with a Nordic accent, and occasional flawed grammar. I guessed she was Swedish, or Danish. We talked for 20 minutes until, when I got up to continue my walk, she also started to pack away.

"You mind if I come to you? Sorry, with you?" she asked.

"Sure."

So far on my adventure, I hadn't walked with anyone, and it felt strange. I have no issues with solitude – I revel in it – but I had spent a lot of time with others on most of my other long-distance hikes. On the Camino I couldn't help but make friends just because of the sheer number of people who walk the route. The same happened on the Pacific Crest and Appalachian Trails.

Elina turned out to be fine company, and over the course of the day towards Kinlochleven we never stopped talking. I remember reading once that a few people come into our lives and remain good friends until we die. Others stay with us for a few months, or a couple of years. But some arrive for a short time to help us, staying around for just a day or two. I didn't realise it but, looking back now, that chance meeting with Elina helped me along another path: my path of depression.

We walked an easy trail towards Kinlochleven. Ferns

speckled the lush grass, which gleamed from the recent rain. Low cloud appeared motionless, obscuring the mountains, doing their best to chase the occasional chink of sunlight away. The WHW crunched under our feet as we weaved around puddles and thudded as we crossed over watercourses on rickety wooden bridges. The wind softened, offering quiet and the chance for contemplative conversation.

"Why are you doing the West Highland Way?" I thought I'd start with the unoriginal, stock hiking question.

"Oh, walking is nice to think," she replied. "Many times, I need to come to nature for a few days and do this. I cannot think at home, or at work. There is too much noise, it's too busy, I go crazy. If it wasn't this route, it would be another somewhere else."

"What are your thoughts?"

"Why I do the job I hate, and why I don't change my life to do something different. Being here, and in nature, my head is clear, I concentrate. I return with new ideas and, I hope, the strength to follow them. All this makes me better."

Pausing for a few seconds, and just as I was about to reply, she added, "also, it helps with my depression."

I looked at her, and she returned my gaze, as if waiting for a reaction.

"I see," was all I could muster.

"Have you heard of depression, Fozzie?"

"Yes. I know it's a common problem, a mental illness. I don't know much about it though." I felt ashamed at my ignorance.

"Do any of your friends suffer with it?"

"My ex-girlfriend had problems with it, yes. She became sad sometimes and struggled to get out of bed. Ended up in hospital. We had problems because I didn't understand it."

We carried on in silence for a few minutes before cresting a hill and stopping. The view took in Loch Leven, its narrow waters softening and decreasing until they vanished into the villages of Kinlochleven and Kinlochmore. The hydroelectric power station was visible – as was my route, which climbed past it, continuing upwards to a col between Stob Mhic Mhartuin and Beinn Bheag. From the col the Way dropped down a mile-long descent known as the Devils' Staircase.

We dipped in and out of woodland. Trees let go of water drops, which fell, smacking our heads and necks. The air was moist and musty, a heady mix of damp ground and decaying leaves. Midges, having sought shelter from the wind amongst the woods, annoyed us as we swiped them away.

"Would you like lunch with me?" Elina asked, running a hand through her wet hair, smiling as if resigned to the fact that being wet and bitten was all part of the Scottish experience. "I stop here for the day, I have to wait for my car back to Fort William."

"Sure," I replied, "I'm starving."

We stopped at the first pub, the Tailrace Inn, and removed our shoes before entering in our socks. The barman smiled in appreciation.

"Thanks for leaving your boots outside," he said. "West Highland Way?"

I nodded, switching my focus between him and the Guinness pump.

"Yup," I replied. "Pint of the black stuff and two menus please. Elina, what are you drinking? The Guinness is good."

"I don't usually drink alcohol," she replied with a resigned smile. "An orange juice is fine, thanks."

We took a table by the window, taking advantage of the mid-afternoon sunshine, which peeked sporadically from behind the clouds.

"You said you didn't understand the depression with your

girlfriend. Do you understand it now?" She carried on the conversation from earlier as if there'd been no interruption.

"A little," I replied. "I did some research after our relationship ended, but when I realised we were finished for good, I lost interest. Perhaps you can explain it?"

"Yes. Depression is a mental illness – this fact is important. Not a physical thing like breaking a leg and solving this by putting it in plaster. It's a condition of the mind, so we know only a little about it. It's hard to explain, Fozzie, and to understand too, but for me, there are many mood swings. One day I'm on top, then bad the next day. I get low, really low. So low I cry."

"Is there anything you can do to lift yourself, to return back to a good place?" I asked.

"Some things help, others don't, but it's not that simple. Depression isn't voluntary, we have no choice. It doesn't care when it hurts me, or how severely. If it decides to give me a bad time, I must let it. I have some control how often it happens, and how badly. For example, I used to drink, now only rarely. It is thought that alcohol makes it worse. This was hard for me because I did love alcohol, although sometimes I do have one. But, *but*," she stared aimlessly out the window searching for the right words, "in the evening it covered the pain, numbed it, and for a few hours I felt better. Then, the day after it just made me weak. It took me two years to admit that drink wasn't helping, so I don't do it often now."

The landlord startled me. "Would you like to order now?" he asked, looking expectantly.

"Sorry," I offered, "we haven't looked at the menus."

"I know what I'd like, Fozzie," Elina said, and ordered a cheese ploughman's.

"Er, do you have any haggis?" I enquired.

"Aye. Neeps and tatties with it?"

"That would be great, thanks."

"Another pint, sir?"

"Yes please?" I said sheepishly, turning my attention back to Elina.

"What else did you do to help?"

"I made lifestyle changes. I used to smoke marijuana, I ate poorly, and I didn't exercise. These things I also changed. I still smoke weed occasionally. I improved my diet and I now eat lots of vegetables. I work out every day in some form. My life was so low at one point I had to change, I had no choice but to try."

"Smoking weed made your depression worse?"

"Yes. Not for everyone, some people it helps, but it's connected. Do you smoke it?"

"Er." I shifted awkwardly in my seat.

"Sorry, I pry."

"No, it's fine. Yes, occasionally."

The landlord returned. "Haggis, neeps and tatties, sir, and one cheese ploughman's, miss, enjoy."

We were both starving; the food was welcome. Elina ate carefully, savouring each mouthful. She cast occasional looks my way, and smiled, seeming content. I couldn't imagine this woman had suffered so much, for she looked healthy. Her hair shone, her complexion glowed, and she appeared at peace.

Still curious, I returned to the conversation.

"Did these changes you made to your life help?" I asked, finishing my last mouthful and sitting back in my seat, stretching my legs out.

"Yes, very much so. It's not that I never have a drink, or smoke weed, have a doughnut, or sometimes don't exercise for a weekend. But I'm strict and attempt to do everything I should, the changes I made I try very hard to keep to. Fozzie, you seem very interested, why is this?"

I paused, deciding whether to tell her, or just let my thoughts slide away.

"The things you said. They sound familiar." I looked at her, feeling ashamed – which my expression must have relayed.

"What things?" She joined her hands and rested her chin on them, raising her eyebrows, making it difficult for me not to respond.

"Well, my trip so far has been great, but there have been days when I felt as though I could have just gone back home. Low moods that came from nowhere. I couldn't get out of them, like I was resigned to my fate. Then, the next day I'd feel fantastic. It's happened a few times on this hike."

"How do you view the future?"

I thought this was a strange question and frowned, perplexed, but answered honestly.

"Sometimes good, other times with little hope. The future looks great one day, and I can see everything slotting into place just how I'd planned. Then, I see it with negative thoughts, and my plans and goals seem unobtainable."

A taxi pulled up, and she got up.

"Fozzie, I could talk to you all day, but I have to go back to Fort William."

"Why didn't you stay here?"

"I couldn't find anywhere with a room so booked for another night at the hotel. I'll get a taxi back here in the morning and continue. What are you doing?"

"I'll do a few more miles then camp where I can," I replied.

"Walk slowly tomorrow, I might catch you."

With that she placed a hand on my arm, smiled, kissed my cheek and left.

I plodded off in a sombre mood, the conversation echoing. Words repeated themselves as if warning me I should remember them, gentle reminders to recall later. Shaking my head whilst attempting to think of something different, I reached a river and sat on a picnic table. The current crept past, the sound of the water barely perceptible.

Depression isn't voluntary, we have no choice.

If it decides to give me a bad day, I must let it.

Alcohol makes depression worse.

How do you view the future?

I shook my head again, agitated at the questions I couldn't shake, and stormed off angrily. After a mile I had calmed and began to meditate. I do this regularly, especially as I walk. It stops the mind noise and assists in clearing interference, to the point where I can almost empty it of thoughts.

I climbed, passing a sign for the Blackwater Hostel. As tempting as it sounded, the sky appeared unthreatening and the forecast promised a spell of dry weather, so I continued towards the col before the Devil's Staircase. I crossed several bridges spanning tumbling burns. The woods thinned, revealing my route ahead before releasing me back into the open. I squinted, as if released from a dark room, as sunlight raced over the mountains, flashing brilliant.

I clipped the finger stretching from the summit of Meall Ruigh a' Bhricleathaid and watched as the track undulated, sweeping left and right, rising and falling as it weaved a route as best it could towards Milngavie, 78 miles distant.

When I crested the final section, and the start of the Devil's Staircase, I removed my pack and sat to take in the view over Glen Coe. Below me, the A82 – the road I had travelled along to Fort William so many times – snaked through the landscape. The River Coupall tumbled out of

Buachaille Etive Mor's massive bulk, gaining impressively the further it progressed as it collected run-off. A tiny patch of snow, way past its sell-by date, clung precariously to the upper reaches of Stob Dearg, the highest point on the Buachaille, Glen Coe's most famous mountain. As I looked south-east, Rannoch Moor glistened in the early evening sun as light bounced around its many lochans. I realised the Devil's Staircase was named because of its testing steepness but thought that Scenic Staircase would be a more befitting name, taking the spectacle into account.

It was 7pm, and I was tired after 18 miles and needed a camp spot. The col offered little as rock blistered the surface. I hoisted my pack and descended the Staircase towards the A82, eager to find something before I reached it, for I knew the route followed the road for around three miles, offering nothing more than traffic noise for a night's sleep.

Spotting tents, I slowed my pace. Four shelters squeezed into the only flat spot available. Pausing, I glanced over as one camper shrugged his shoulders, as if realising my fruitless search was to continue. I reached the bottom, my legs running on reserve, and resigned myself to a night by the road.

A track on the opposite side led to the River Coupall. I walked a way down it, but a cottage looked occupied and a car was driving around. Not wanting to disturb anyone I crossed back over the road and continued. The traffic was light, the noise unobtrusive. I continued until a rocky outcrop shielded me from passing motorists, and I set up camp.

I pitched my tent, inflated my mat, and left my sleeping bag on top to fluff up after its incarceration in its stuff sack. I filled a pan, lit the burner, and began rummaging around my food bag to find suitable nourishment. Eighteen miles is not far for me, so I was surprised how weary I was and decided sustenance, or rather lack of it, must be to blame. Extra portions should help; I remembered a beef stew I had picked up in Fort William. I read the instructions, which stated serving size was for two people, and therefore perfect.

"Smells great!" called a jogger as he ran past.

"Enjoy!" cried his companion a few seconds later.

Elina's words, and our conversation, continued to resonate. I replayed it, not becoming stressed as I had done by the river in Kinlochleven. Thinking about my three weeks in Scotland, and my state of mind, I realised her description of depression shared similarities, but I dismissed them.

You're not suffering with depression, Foz, don't be bloody stupid.

Yet what she'd explained seemed familiar. The mood swings, those low days when everything was just too much effort, and the day after when the world was glorious once more, leaving me confused how my emotions could change so quickly.

I thought back to the pit. How, one minute I was hiking without a care and happy, the next staring at those cold, dank walls wondering if I'd ever escape. I mulled over my drinking, and that every time I was near a pub in the evening I ended up going inside and having too much. The days after those drinking sessions were often emotional. The pints and whiskies at Strathcarron, and my further inebriation in the pub that day in Inverness, left me down the following day.

I took another swig of Scotch before falling asleep.

Chapter 9

The Valley of Happiness

Fish and chips are a great mix, gin and tonic are best buddies, and Laurel and Hardy a classic combination. Most people would consider being introverted, depressed and solitary an unfortunate mix. On its own, each trait is misunderstood and stigmatised – put them together and it doesn't bode well, does it?

Consider this. If potential dinner guests had to fill in a CV and apply for the dining position, then an introverted, solitary depressive would have little chance of even making the shortlist. If accepted, you'd be eating in the back garden, patio doors locked, while your entrées and the occasional drinks refill were passed out the window.

No! Don't talk to him! He's depressed. As if that's not enough, he's introverted and prefers his own company. Imagine! He wouldn't even know what to say!

If you met me, it's unlikely you'd be aware I'm any of the above. I'm articulate, I don't sit in a corner hiding under a table, nor do I have a dark cloud hovering above my head –

although on the rare occasions I do go to a party, I hang out in the kitchen.

Despite appearing comfortable, I hate social events and have done for years. I hide it well enough, but there comes a point when there's only so many invitations a person can decline. Believe me, I've used all the illnesses and excuses I can.

I love being introverted as I do being solitary. Neither bothers me, as I like my own company, and I'm not dependent on being around others for stimulation. A few years back, more the wiser for age, and less bothered about what people thought of me, I told my friends how it was. I explained my introversion, that I found socialising difficult and draining and preferred not to be there. Now they have accepted this. I'm not saying I decline everything; some days I'm more confident than others.

In the months after my hike, when diagnosed with depression, I realised that I wasn't comfortable with it. The lack of control was the frightening part. I never knew when it was coming, how long it might last, or how severely I'd be affected. Unable to mount a challenge, my actions felt inadequate, despite my best efforts.

The fact is we all have our issues; we all battle our own monsters. You have no idea that your friend, the company executive with the loving wife and two adorable kids, has a terminal condition he hasn't told anyone about. I don't know if people around me have bipolar disorder or have an incurable disease. Others may be in serious financial trouble, on the brink of bankruptcy, or contemplating ending their lives.

I gain no comfort from the thought that others may be faring worse than me, but it puts everything into perspective. During the horrible days, the times when I, too, thought about suicide, I realised that millions of other individuals around the world were facing harsher problems and winning their own battles.

If they could defeat their demons, so could I.

Scotland has more than its fair share of atmospheric locations. Its bloodied history, legends, and tales passed through families over centuries, and the magnificence of its geography, enrich this corner of Great Britain. Perhaps it's the spirits of those lost souls that continue to linger in the trees, rocks, and rivers. Maybe they still roam, invisible to us, but I swear these places don't forget their past. They make the hairs on my arms stick up.

Glen Coe is such a place.

I forgive the busy A82 running straight through the middle. For those who don't often experience the outdoors, the road gives them a glimpse of one of the finest glens as they drive past. Much of Scotland's scenery is awesome, so wherever you build a road, it's likely to intrude.

I've stood in many scenic locations around the world, motionless, just taking them in. As much as I realise how mountains look, how calming the depths of a forest can be, how clouds play games with the sun, casting shadows miles long that race along, up and over everything in their path, many spots appear surreal. Despite witnessing rivers carve through undefeatable obstacles, regardless of how immense valleys rise and leave me spellbound, I continue to be humbled.

Scotland is up there with the finest, and if you don't enjoy the outdoors but do drive along the A82 through Glen

Coe, treat yourself. Stop the car, walk a few hundred feet from the road and sit for a while.

I slept well, despite the early sunrise illuminating my tent. Then I realised.

Hang on a sec, the bloody sun's out!

At the end of my trip, out of curiosity (and slight resentment), I calculated how often it had rained. Of the 31 days I took to traverse Scotland, 29 of them had been wet. This is not conducive to a healthy state of mind. As Nick Levy, a friend with whom I hiked part of the Pacific Crest Trail, said: "You can push through many obstacles with determination, but the hardest is poor weather. It's belligerent – it will break you eventually."

I unzipped the tent and peered out, squinting at the intense light bouncing around Glen Coe. I escaped from my sleeping bag and stood to soak it up.

Oh, you got to be kidding me! This is fantastic!

At that precise moment I remembered I hadn't spent the night somewhere remote and hidden from view. In fact, I was 100 feet from one of the busiest roads in Scotland, and stark naked.

As I suddenly realised this fact I caught sight of a woman speeding past in the passenger seat. I followed her eyes as they met mine; they moved downwards and paused, longer than was necessary. She screamed with surprised laughter before turning to relay the news to the driver, and then the car vanished behind a hillock. My hands dropped as I dived back in the tent to put my clothes on.

Scotland had changed dramatically. The sky was an intense, cloudless blue. Huge banks of mist floated through

Glen Coe, creeping between Stob Dearg over the far side of the valley and Beinn a' Chrulaiste behind – spectres riding the weakest of breezes, passing and eyeing me cautiously. The temperature dropped a few degrees as these spirits passed, then rose as the sunlight broke through again. Like theatre curtains opening and closing for several, well-deserved encores, Scotland vanished then reappeared, bowing. The ghosts of Glen Coe themselves had engulfed me.

I made coffee, then tipped dried milk powder into my pot with water and stirred, added granola, crumbled in some dark chocolate, and drizzled it with honey. Sipping my drink and munching my breakfast, I nodded the occasional greeting to hikers, who seemed surprised I had camped by the side of the road.

"Fozzie! I caught you!"

I turned my head in surprise, shocked that Elina had arrived so fast.

"What are you doing here so quick?" I cried from a mouthful of granola. "Sorry, I have food in my mouth!"

"I can see!" She removed her pack and sat beside me, rummaging around for a snack. "Well it is 10.30."

I checked the time in disbelief; having only been up 30 minutes, I must have woken at 10.00. I'd slept for 12 hours straight. Before I could fathom how, she continued.

"The taxi dropped me back to Kinlochleven at 8.30. I've been walking since then. Why did you camp here? What about the cars?"

"I got off the Staircase late, and the trail sticks to the road for a while yet. I was tired, so I stayed here. Not ideal but surprisingly quiet."

"Look at this sun, it is warm, and bright! So bright!" She screamed in delight.

I watched as she strolled, pausing to hold her face up, arms outstretched. She looked elated, and for a second, ashamedly, I was jealous. It was simple; she was happy, I wasn't. How had this girl overcome her demons and escaped her torment to be the woman I saw now? I suppressed a feeling of unfair resentment and pondered how I could liberate my own anxieties.

"Fozzie! Come on! Let's go! I want to walk through Glen Coe!"

I flitted between packing, smearing sunscreen on my face, and trying to find my socks.

"OK! Ready!" I cried, smiling in anticipation. "Let's do it!"

We'd only managed an hour's hiking when the Way Inn appeared.

"I need coffee," Elina declared.

"Me too," I replied. "Oh, look at that, they're doing cooked breakfast."

"You just ate!"

"That was at least an hour ago." I rubbed my stomach. "Besides, firstly, never, ever pass up a cooked breakfast, and second, we should talk."

She raised her eyebrows, appearing confused, and smiled.

We placed our coffees on a table, spoons chinking as sunlight caught the rising steam. Elina sipped on her coffee and sighed in appreciation.

"Do you mind telling me more about your depression?" I asked her. "You don't have to if you don't want to."

"It's fine, Fozzie, I'm happy to. I hadn't been right for a year but thought I was just down a little. There were horrible periods towards the end of that time – over the course of a few days I cried uncontrollably, sometimes shaking for no reason. I didn't want to see anyone or work. I said I had a cold and hid from everything for a week."

She peered out of the window, glanced my way, and smiled. I sensed there was more coming.

"A friend called. I told her everything that was happening. She ordered me to go to the doctors that afternoon and to tell them I might be suffering from depression."

"How did she know?" I asked.

"Because she suffered with the same thing."

"What did the doctor say?"

"He asked me a lot of questions, about sleep, if I drank alcohol or took any drugs. He wanted to know how I felt about life, my future, how I coped with my job and other people. I think these questions gave him an idea of if I had depression, like a checklist. He recommended pills, I forget name."

"Antidepressants?"

"Yes, those. I said no, so he strongly suggested that I stop smoking marijuana, cut my drinking if not stop altogether, and concentrate on looking after myself. He put me in touch with a counsellor. Fozzie, do you think you—"

I cut her off. "No, I'm not depressed, just down that's all."

"There's no shame in it. If you're unsure, see a doctor."

"Yeah, OK."

We finished our bacon rolls, walked out into the Scottish sunlight, and carried on south. As if the topic had been resolved, we talked about other things. The path was busier; hikers passed every few minutes nodding, smiling, and offering greetings. They were happy, out enjoying Scotland in beautiful weather.

It was strange. My situation couldn't have been more different a few days earlier. I thought back to when I'd left the Cluanie Inn, climbed into the clouds, walked in rain for most of the day, crossed rivers, and saw just one person. Now, I had a companion and conversation, I'd passed many hikers, I was wearing shorts and a T-shirt, and there wasn't a cloud in the sky. Instead of wanting the trail to be over, I hoped it would last forever. It was limitless Guinness, but I wanted endless mileage as well.

Although curious about the discussion with Elina, I believed I was OK. In hindsight, if I'd digested what she had been telling me, and followed the trail of clues, I might have realised earlier.

It's hard to explain, Fozzie, and to understand too, but for me, there are many mood swings.

I get low, really low. So low that I cry.

Depression isn't voluntary, we have no choice.

There's no shame in it.

Having been surrounded by glorious Scottish peaks for three weeks, I drifted back to memories of my limited (and hesitant) dabbling in mountaineering. I hadn't taken to chasing serious height via glaciers, snow chutes, and ice faces. I enjoyed elevation, and reaching summits, but I preferred the easier routes to get there. Ropes, harnesses, helmets, and crampons felt like complications. However, my main issue centred on the risk – and being unable to come to terms with it. I understood and sometimes enjoyed the danger buzz, but I could never shake the uncomfortable sensation that came with it. Up high I felt close to death.

Years earlier, during a brief spell when I had ventured into the sport, I travelled up to Scotland with a mate, Jeremy. We hired an experienced guide for a week, Zac

Poulton, to gain more winter mountaineering experience. Before, I'd built my winter knowledge in the Lake District, visiting classic locations such as Striding Edge, Scafell Pike, and Great Gable.

It was a great few days, the weather warm for mid-February, and one morning we'd stripped to T-shirts. A mate of Zac's, Christopher Walker, acted as assistant guide.

The day centred on the ascent of Bidean nam Bian, a lovely peak of 3,773 feet overlooking Glen Coe. The highlight for me wasn't the mountain but the chance to walk through Coire Gabhail, more commonly known as the Lost Valley. Parking near the A82, we thudded over a bridge spanning the River Coe. After passing through a copse and scrambling over a mishmash of fallen boulders, the trees opened out, the rockfall fell behind me, and I finally entered the Lost Valley.

I'd heard about the place from a friend, who implored me to visit, and now I understood why. Concealed from the road, and only revealing itself at the last moment, it appeared as if hidden from time itself. A giant colosseum circled by peaks surrounded me.

We worked our way through the valley to the headwall leading up towards Bidean nam Bian's summit, gazing up at the menacing ascent. A steep snow slope topping out to a cornice high above greeted my gaze. We put on helmets, donned crampons, and grabbed our ice axes. Zac offered me the chance to lead, which I accepted, having more faith in my fitness to climb hard than my knowledge of how. I was scared, but buzzing with it.

The gradient steepened near the top, and Zac took over before pausing at the overhanging cornice above him. He scratched his head before tunnelling upwards and breaking out onto the ridge above. The rest of us followed. The remaining ascent was a leisurely plod along a wide ridge to the summit before we began our descent.

Zac and Chris sized up route options on the way down, sharing their avalanche knowledge as we listened and observed. They both discounted one descent, as they calculated the risk was too high, and we safely returned via another route. I remember being impressed by their acumen and grateful for their leadership.

Two years later, on Buachaille Etive Mor, Christopher was guiding two clients, Ritchie Birkett and Robert Pritchard. He dismissed the normal descent into Coire na Tulaich and chose to descend the ridge, which he thought was a safer choice. Ritchie watched in horror as Christopher and Robert descended, only for the ground in front of them to avalanche, taking both climbers with it. They both died in the accident.

I was shocked when I found out. It struck fear into me, first because I'd met Chris, and second because it was so close to where we had climbed. The sheer number of factors a mountaineer has to deliberate is staggering, and, despite what they perceive as the best decisions, despite best intentions, sometimes it goes wrong. This is the unknown that scares me.

Three years after that Scottish trip, I drove to the Swiss Alps with Jeremy for a week of more advanced mountaineering tuition with Rich Cross and Olly Allen from Alpine Guides.

Our first Alpine peak was the Allalinhorn, popular for two reasons. First, it's 4,000 metres high (4,027 to be exact), so a big draw to those wishing to add kudos to their mountaineering experience. Second, a cable car ascends to the Mittelallalin top station at 3,500 metres, leaving an easy climb to the summit. This limits a climber's chance of suffering from acute mountain sickness. Everyone is different, but the risk starts around 2,500 metres and can cause dizziness, nausea, headaches, and shortness of breath. Taking the cable car, bagging the Allalinhorn, and getting back to a lower elevation was a great idea.

It started well. The weather was perfect although storm clouds were bubbling to the west. Rich, Olly, Jeremy, a few other clients, and I made excellent progress up the steep but manageable approach over glaciated terrain. Arriving at the Feejoch col at 3,826 metres, we rested before turning east.

The elements had deteriorated, and dark clouds rumbled. The proper course of action around electrical storms up high is simple: you descend. Rich was ahead and out of sight, having gone to check on the situation. I was chatting to another member of the team, Roger, when a thunderclap from hell itself cracked above us, and we cowered like frightened kittens. Seconds later Rich came sprinting into view, shouting and waving his arms frantically.

"Get down!" he screamed. "Go back now! Go!"

Olly had already provided lessons in crampon use, advising we space our legs further apart than usual to prevent the spikes clashing and causing trips. His favourite suggestion was to 'walk like John Wayne'.

I was panic-stricken, but, looking back, I can think of few funnier memories. We turned and bolted without having had time to remove our crampons. Fleeing the Allalinhorn in full mountain gear, on icy slopes with crampons, could never be described as graceful. Having descended to a safe elevation, we could do little but laugh at our misfortune and failure to bag our first 4,000er.

Towards the end of the week we attempted to climb the Pointes de Mourti at 3,563 metres. It had the elements of danger one might associate with a mountain: the crossing of the Glacier de Moiry and its associated crevasses, tackling a knife-edge ridge, and drops so high that my chocolate Nesquick breakfast was curdling. We had stayed overnight in the Cabane de Moiry, an Alpine hut at 2,528 metres, and made an early start at 3am – a common strategy in alpinism to make use of the colder and firmer snow.

As the morning progressed, my nerves frayed. An hour into the route, we passed under an ice face that cracked and groaned, seemingly ready to detach.

Falling from a height or drowning are the two most frightening ways I can think of to die. When confronted by crevasses on the Glacier de Moiry, I couldn't stop thinking about this, because some crevasses reach flowing water deep under the ice. If I fell into one, two of my worst fears could be ticked off in one, plummeting moment of horror.

Olly led, roped to Jeremy in the middle, while I followed up at the rear. Our guide picked a careful route through a crevasse maze. Ice cracked beneath us. As I looked up in between concentrating on my foot placement, Jeremy suddenly vanished.

"Olly!" I cried. "Olly! Jeremy disappeared!"

'Jeremy disappeared' wasn't the best description I could have come up with, but when I finished dashing to the spot I had last seen him, it was appropriate.

"You OK?" I asked, looking down at my best mate's head, all that was visible as the rest of his body dangled inside a crevasse.

"Yeah," he replied, giggling.

Olly arrived. "He looks like he's been decapitated."

We pulled him out, at which point Rich arrived with two other members of the group and informed us we'd be practising crevasse rescue. My heart sank. Presumably, this entailed someone having to get into a crevasse to be rescued, which I hoped wouldn't be me.

"Fozzie," Rich announced an hour later as we stood by the side of a gaping crack, "you're first."

"I am?"

"Yes, clip your harness in, the rope will hold you. Don't worry, we got you."

I was about to protest, alarmed at the prospect of getting up close and personal with my worst nightmare, when Rich handed me the rope.

"You'll be fine, mate," Jeremy chipped in, seeing the look of horror on my face.

I took the rope, clipped it to my karabiner, and screwed the gate shut. I checked my harness, the rope, and the karabiner again in a desperate attempt to delay my incarceration. Then I looked at Rich, just in case I had misheard him. I hadn't.

Inching closer to the abyss like a frightened child, I glanced upwards at Jeremy, who smiled, a little too smugly for my liking. As the edge steepened I couldn't hold myself any longer and slid. But when the rope tightened and my fall arrested, I took stock.

My main fear with mountaineering is ropes, harness, and other paraphernalia. I'm petrified of equipment failure. I looked around, dangling in an ice box. Slowly, my dread subsided. What greeted me was unexpected, and beautiful: the top layer of snow merged into turquoise ice, speckled with air bubbles. As I looked down, turquoise melted into darker shades, deepening to the richest, inky blue I'd ever seen. The light faded below as darkness crept up from the depths.

Despite only dangling 15 feet from the surface, it was quiet, save the occasional drip of water and groaning from the glacier. As you'd expect from being ensconced in an ice box, it was also extremely cold.

I heard Jeremy's voice slide over the crevasse lip, giggling.

"We could leave him a bit longer."

"No, you bloody can't!" I cried.

I felt my harness tighten as the others hauled me out and I escaped my ice prison into the sunlight.

The final test of my mountaineering sanity came near the summit. Olly was leading the final push roped to me, with Jeremy last in the line. Looking around, I realised not only how high we were, but how exposed. The ridge was broad but culminated in a vertical drop off to my left. To the right a slope steepened to disappear down to the village of Les Haudères, a few thousand feet below. We were wearing crampons; although walking on snow, it was hard, slippery and none too stable, my spikes piercing the surface just enough to give a firm footing. As we ascended my fear grew. Olly was positive, encouraging, and relaxed. He appeared as if on a Sunday morning stroll to the pub. I turned to check on Jeremy, who also seemed untroubled. My main concern was the drop to either side.

"Hey Olly!" I cried.

"Yes Fozzie?" he replied, not breaking his stride.

"What happens if you fall over that drop on the left? What I am supposed to do?"

I knew the answer but sought reassurance.

Olly was silent for a few seconds before turning and smiling.

"You jump off the right to counterbalance my fall. Don't even think about it, just do it."

We made the summit and got back down, but Olly's words rang in my head for weeks after. Coupled with the memory of Chris Walker's death, that signalled my decision to stop chasing mountaineering.

Elina and I headed towards the Glen Coe Mountain Resort before veering left, threading a route between the grass-coated Beinn Chaorach and Creag an Fhirich. Many of Glen Coe's finest mountains peeked over Creag an Fhirich; some I remembered from my trip with Jeremy. Clipping the edge of a forest, we crossed a bridge over the River Ba, marvelling at the wide mountain scenery surrounding us as a breeze threatened to move on a few isolated clouds. We were now walking through the vast expanse of Rannoch Moor, the wildest section of the trail.

My expectations of the West Highland Way were correct. On my second day I was enjoying the humble novelty of dry feet. One of many positives gained from outdoor trips is how the little details mean much more.

"My feet are dry!" I proclaimed.

"You are sounding surprised?" Elina replied.

"Believe me, you have no idea!"

The track was a good six feet wide with a smooth surface, as if the maintenance crews regularly ran rakes through the top layer. The grey path, visible ahead and behind, snaked through west Scotland, side to side and up and down. My path hadn't disintegrated into a quagmire once, and the midges had vanished. I relaxed, content that although I still had 311 miles of 600 to go, I had conquered the hardest part.

Skirting the edge of Loch Tulla, we stopped at the Inveroran Hotel. We both ordered tuna melts and more coffee then sat outside with our faces directed upwards to the sun.

"Scottish sunshine has to be taken advantage of to the full," I said with my eyes closed.

"I agree," Elina replied, before continuing, "where are you heading today?"

"Not sure. You?"

"I'm staying at Tyndrum."

Tyndrum was seven miles distant and offered several accommodation options and places to eat. From there, we were 12 miles from Loch Lomond, a section I was looking forward to. I needed a wash but didn't want the expense of a hotel so decided to carry on with Elina to Tyndrum to get a good meal inside me and see if there was a campsite with showers.

One section of the WHW I used to peer at during my drives to Fort William lay between Bridge of Orchy and Tyndrum, where a magnificent valley was visible from the road. The trail hugged a low route next to the Allt Kinglass river, and now I would hike it at last.

We left Bridge of Orchy and passed under a railway bridge, entering the glen. The peaks of Beinn Bhreac-liath to the west and Beinn Dorain to the east marked an impressive boundary, funnelling us towards only one possible outcome, the mighty Beinn Odhar. Standing guard at the valley's head directly in front, Beinn Odhar made it easy to stay on course – and as we moved closer, this beast came up close and personal. Its classic cone shape, all the more impressive as it reared up from the low of the valley, exaggerated all 2,956 feet of its bulk.

That day I was happy, and my happiness focused on that place. I forgot about depression, my ignorance of it, the constant fear of the pit, and the insecurity of when my life could plunge downwards. I remember it now as the valley of happiness, and, although I have not returned to Scotland at the time of writing, I know that my next trip will be different. Instead of looking down from the car, I'll park and walk there to rekindle that memory.

Days of optimism during my adventure were few. After I'd

returned, I realised with sadness that there had been just a handful. I'm not suggesting the remaining time was miserable, or forgettable – it wasn't. At best, I coped. Scotland is mixed memories of confusion then understanding, of frustration then jubilation, dejection and happiness. This mesmerising land of mountains wasn't to blame, and its attempts to heal were valiant, but my rehabilitation didn't start until I returned home.

But, Scotland triumphed in one respect: directing me towards understanding.

We crossed over the Allt Kinglass. Having reached Beinn Odhar, the path, with nowhere to go, ushered us south along the Allt Coire Chailein. We climbed steadily, our breathing laboured and T-shirts soaked with sweat. Resting on an old stone wall, we munched on snacks and spoke little, content with silence, the scenery, and beautiful weather. We passed alongside the delightful former miners' cottages in the hamlet of Clifton and followed a stream through a field dotted with a few tents.

"I'll camp here," I said, "too lovely not to."

"I'm staying in the Tyndrum Inn," Elina replied. "Come for dinner later, around seven?"

"Love to, see you then."

I scouted for a flat spot and settled just by the stream, which gurgled gently, promising a good night's sleep. Others had camped on a prominent position enjoying commanding views and nodded while attending to their grilling meat.

My familiar aroma of hiker filth was increasing, and I'd noticed a tent symbol on the map. I wandered off to find the Pine Trees Campsite.

"Hello," I said, "I don't need to camp, but would it be

possible to use your showers please?"

"Yes, of course," the owner replied. "We have soap and shampoo if you need it."

The water was steaming hot, and I let it run over me for a while, feeling it ease my aching muscles. I returned to my tent and updated my diary, nodding to people out for an evening stroll.

The Tyndrum Inn was an old building with a plastic conservatory bolted on one side, like a deliberate attempt to sabotage its appearance. The interior was typical pub: lots of wood, burgundy carpet, and brass ornaments that caught the light. I found Elina sitting in the corner, reading.

"I thought you didn't drink?" I commented, noticing her hand cupping a glass of wine.

"No, Fozzie, I said I drink occasionally. I felt like some wine."

"My mistake. How's the menu looking?"

"Good. The usual food but everyone is eating happily." She smiled in a resigned way as if something had happened that she had little choice in.

"I have to go tomorrow," she added suddenly.

I sat. "You do? Why?"

"Family problem. A relative is ill, and I need to get back. My flight leaves midday. It's sudden, I'm sorry."

"You have nothing to apologise for," I said. "I hope it works out. I'll see you before I set off in the morning?"

"Yes, I'd like that, to say goodbye."

It sounded so final. We only had three days to Milngavie, the end of the WHW, and I'd never even expected to meet Elina, or walk most of the way with her. But the thought of being back alone, despite my love of solitude, was suddenly frightening. I felt scared, like on my first day of school.

We spent the evening chatting, enjoyed great food, and

joined in with the locals who were short of members for the local darts team, before going our separate ways. At ten o'clock, as I reached my little glen, the sun was setting. Startling reds and oranges streaked the heavens over a forest as if the trees themselves were ablaze. The rest of the night sky faded from this scarlet fire and gradually turned black as if extinguished.

I sat for an hour outside my tent, listening to the stream and watching the colours glow. I dreaded how I'd feel in the morning. The weather was due to turn again, and my companion was leaving: two minor events that I'd have normally dealt with but, coupled with a series of wonderful days up to that point, the fear of change was overwhelming. Despite the stream's efforts to lull me away, I slept poorly.

I returned to the Tyndrum Inn in the morning for breakfast with Elina. The expectation of parting ways led to a sombre hour before her taxi arrived.

"Be happy, Fozzie," she said, kissing my cheek before placing both hands gently on my face, looking straight at me, concern evident in her eyes. "Remember to be happy."

"I will. Why are you saying this?"

"You will find out later, just... just be happy."

And she left.

I walked over to the local store to buy food for the day, then set off with high expectations of reaching one part of the trail I'd been looking forward to: Loch Lomond. Despite Elina leaving, I did my best to stay upbeat. A low-slung mist once more shrouded Scotland, but the sun soon burnt it away, revealing the countryside. Entering forest, I gained and then lost height repeatedly as the track rose and fell. Occasional breaks in the trees revealed glimpses of the valley below, where the River Fillan shimmered.

The area around Tyndrum has a long history of mining,

and gold is still extracted there now, the remnants of the industry evident as I progressed south.

Legend says that after Tyndrum, near Strathfillan, Robert the Bruce had been pursued by the English army, under the command of Aymer de Valence. Bruce had fled, having refused to swear allegiance to Edward I of England, only to find his escape blocked by a thousand MacDougall soldiers – who also had problems with Bruce after he had killed another Scottish noble, John Comyn.

Bruce had no choice but to stand and fight. Somehow, despite being outflanked and heavily outnumbered, he fought for his life and won. Afterwards he fled with a handful of his men, and to aid their escape he ordered them to throw their weapons into Lochan nan Arm to lighten their load. It's thought his legendary sword still lies somewhere in the water's depths.

I reached a shop called the Trading Post. Although not intending to stop, a man exited clutching a cup of coffee, encouraging me to do the same. As I sat outside I noticed a piece of folded paper in the side of my pack and pulled it out. I didn't recognise it, nor did I remember putting it there. As I unfolded it, a handwritten note met my gaze.

Dear Fozzie.

Excuse my poor English writing but I could not have told you this to your face.

Either you hide your emotions well, or you do not realise what battles you are fighting. I see in you things I saw in me a few years ago. Frightening things that I have now fought, and controlled, but I hope you understand that I have to tell you what these things are, so you too can do something.

When you have depression, you can see this in other

people. I see signs, and from talking to others, can recognise from their words about their experiences that they may, also, have this illness.

I see this in you.

Please forgive my honesty, I hope you understand. I had a very special few days with you, and I hope we see each other again.

For now, you must finish your walk. Use the time to think, and to answer the question of if you are happy. I think from your words that the answer is no.

Finish, return home and please see a doctor, this is the first step you must take.

Remember, be happy,

Elina x

My eyes welled with tears as I read it a second time. I thought it impossible and shrugged off the notion I had depression, even though I was just realising what it was. I yearned to talk to her again, and the feeling of loneliness returned. What the hell was wrong with me? Maybe Elina was right. As she said, she was qualified enough to recognise the signs. I pondered for a few minutes while I finished my coffee, then set off before my negativity deepened.

Cresting a hill near Inverarnan, I stopped dead. Before me stood Loch Lomond, sparkling gloriously in the midsummer sun, stretching 24 miles south through the Loch Lomond and the Trossachs National Park. The WHW clings to its eastern edge. It would take me an entire day from reaching its northern tip to leaving at the southern shore the following day.

Shaded by woodland for most of the way, it made for lovely walking. The first part caught me out. I was expecting the classic, flat trail I had become accustomed to, but the track went crazy as it tried to hug the loch edge. Stob nan Eighrach and Beinn a' Choin lifted skyward to the east, angling down to lunge into the loch's depths, creating a bumpy ride. Clambering over fallen trees, up and over dirt banks, treading carefully across sections of tree roots, I tired quickly. Other hikers coming from the south wiped ruddy brows, sweating and looking surprised.

"It's crazy, no?" a French woman called, giggling and wiping a bloodied knee.

"Oui!" I cried.

Gradually, my path settled to a level, smooth track. Trees climbed the slopes to my left as the clear waters of Lomond flashed in the sunlight to my right. The loch barely registered a ripple, a silent, dark abyss plunging 620 feet at its deepest point. I rested on a boulder by the edge, tossing small pebbles, watching the surface react and undulate until the energy dissipated. Boats passed, and I observed their wakes travel closer, lapping at the shore, the merest hint of sound confirming their arrival.

I saw Island I Vow from my vantage point. Three hundred feet long but poking only 30 feet above the water, this small lump of rock was once home to the MacFarlane Clan, and the ruins of their castle are still visible. It was built to replace their castle on Inveruglas Isle, further down the loch, destroyed by Cromwell's Roundhead troops in the 17th century. A hermit inhabited I Vow in the 1940s, living in the castle dungeon. He marked paths using pebbles from the beach, and they still exist today.

Early evening, having completed a third of my journey along Lomond, I arrived at the Inversnaid Hotel. A solitary road connects the grand building with Aberfoyle, 15 miles distant. Built in 1790 for the Duke of Montrose as a hunting lodge, it surprised me as it appeared from nowhere, seeming out of place.

Despite my resupply in Tyndrum, I had neglected to buy food for dinner, so I entered in search of sustenance and ordered a beer at the bar.

"Presumably you're doing food?" I asked the barman.

"We are, but you may be out of luck I'm afraid," he said. "We have one sitting for the guests, and I believe they're full in there. Hold on though," he added, "I'll ask."

He walked over to the dining room entrance and held a brief conversation with a waitress before returning.

"They have room for you, yes. It's three courses, I can take the £10 payment now for you?"

I paid him and joined the long queue stretching back into the corridor. As we filtered into the dining room, I felt as though I were back in a Butlins holiday camp from my youth. I'd guess there were 200 people seated on rows of long tables. The entire serving staff lined up, sporting matching black attire, nodding, smiling, and saying hello to everyone as we entered.

I acknowledged my fellow guests and took a seat. The menu had three options each for starter, main course, and dessert. I ordered vegetable soup, pasta in tomato sauce, and fruit. A few minutes later the kitchen doors swung open, and an impressive line of servers emerged holding our starters, like one long, black caterpillar. This was a highly tuned military operation: each bowl was placed on the table, before the servers returned to the kitchen and appeared again with more dishes. At one point I swear the serving staff outnumbered the guests.

The caterpillar returned 15 minutes later, weaving around tables taking finished bowls in exchange for the main course, and the process repeated for dessert. The food was decent enough, especially as it only cost me a tenner, including a glass of wine.

As I chatted to Bob from Middlesbrough, on his yearly fishing trip, we couldn't help but calculate the total bill, and the overheads. We figured they just about broke even.

As an encore, while we sat chatting over coffee, the entire serving staff lined up once more, bowing as we clapped in appreciation. I've never witnessed a meal like it.

I retired outside and looked around for somewhere to camp while smoking the last of my blueberry crush. Much of the eastern shore of Loch Lomond is now subject to strict laws that forbid camping. Fines are in place to enforce them. My trip was before these restrictions came into effect, but many have since fiercely contested them as contravening the basic rights of wild camping in Scotland.

I found an idyllic location right on the loch shore and watched clouds of midges collect on the tent netting, making vain, frustrated attempts to reach me. I read Elina's letter once more, the contents repeating. And I repeated more words before I fell asleep.

I'm fine. I am really. I'm not depressed.

Chapter 10

By the Canal

I was away by 7am, sinking deep into my jacket to escape an unwelcome chill, despite cheerful birdsong reminding me it was midsummer. I dawdled, enjoying the peace of early morning.

If we are to believe the geologists, Scotland was once part of the United States. This was so long ago it's difficult to picture. However, not only was it easy for me to imagine, the similarities were obvious.

The Appalachian Trail had been on my mind for days, and nothing seemed to have changed since the separation of the continents.

During my time spent along Loch Lomond, I could have been on the Appalachian Trail. The resemblance between that corner of Scotland and the north-eastern American states was undeniable.

The forests in Scotland mirrored my memories from Maine and other northern states such as Vermont and New York. Even the lochs, cupped into valleys, their water almost black, were strikingly similar. In shallower water near the shore, the black lightened to a peaty brown that tainted the stones and pebbles underneath into shades of bronze, copper, and gold.

The trail occasionally revealed the dark soil through a carpet of pine needles, and rocks poked out of the ground as if coming up for air. The dramatic landscape which tumbled from up high, plummeting and rolling for miles distant, was similar to Maine.

Looking back to my time on the Appalachian Trail, I could have been in Scotland. And, during the latter stages of the West Highland Way, I was back on the Appalachian Trail.

The International Appalachian Trail organisation (IAT) is taking this shared history to the next level. The Appalachian Trail runs from Springer Mountain, in the southern state of Georgia, to Mount Katahdin in Maine, a distance of 2,181 miles.

250 million years ago, the Earth's plates collided, creating the supercontinent Pangea and forming the Appalachian-Caledonian chain of mountains. Over millions of years, Pangea drifted apart to form the Earth as we now know it, and the Appalachian-Caledonians split, redistributing to North America, Greenland, Western Europe and north-west Africa.

In 1994, Joseph Brennan, former governor of Maine, proposed extending the Appalachian Trail north from Katahdin to Mont Jacques Cartier in Quebec's Gaspé Peninsula. The beginnings of the IAT were born.

Over several years, plans to lengthen the IAT were

proposed. Delegates travelled to other nations to explore the possibility of bringing other countries on board. In 1997, John Brinda became the first person to hike the entire IAT as it existed then – and the entire east coast of the North American continent – in his trek from Key West, Florida to Cap Gaspé, Quebec.

Interest soon turned to the eastern side of the Atlantic and the possibility of utilising existing trails in the UK, France, and in North Africa. Progress has been remarkable: at the time of writing, routes link the IAT through Maine, New Brunswick, Quebec, Nova Scotia, Prince Edward Island, Newfoundland, Greenland, Iceland, Norway, Sweden, Denmark, Northern Ireland, Ireland, the Isle of Man, Scotland, Wales, England, France, Spain, Portugal, and Morocco.

In Scotland it encompasses 492 miles – 217 on the Cape Wrath Trail, 96 on the West Highland Way, and 179 on the Firth of Clyde Rotary Trail.

Currently, the IAT is around 12,000 miles long. The route still needs to be finalised in Spain, Portugal, Scandinavia, and France.

That's some hike!

I'd covered a mile when a delightful cottage appeared, but it was the setup outside that lured me: a cabinet, stuffed with homemade treats by the residents, who had lovingly crafted chalkboard signs to advertise the bounty. It said:

A wee treat to help along the way.

Granola bars and cookies were crammed into various plastic containers. I scoffed one of each and left a donation. A large glass vat at the top, full of water, infused with lemon and orange slices bobbing around inside, washed down my impromptu breakfast. I think the owners must make a tidy profit in the summer from the scores of hikers.

I passed the Ben Lomond bunkhouse and the Rowardennan Youth Hostel, losing count of the number of bridges I'd crossed, before reaching the village of Balmaha. I turned, faced Loch Lomond, and saluted a farewell, as I knew the trail now left the water for good.

A long walk by the roadside followed, and, through breaks in the trees, I observed Scotland changing. Behind me, Loch Lomond and the Trossachs stood proud. Ahead, the land relaxed, smoothing out to lower elevations and a less rugged landscape, although still impressive.

Milngavie was close, a day distant, just to the north of Glasgow. I had no intention of progressing through the big city, purely because I wanted to stay in the countryside. I calculated a day of 28 miles, leaving a short 12-miler the following day to Milngavie, where I planned to splash out on a nice hotel. This distance meant I could arrive early and make the most of the room – a practice I used regularly when hiking. I needed to do laundry, shower, buy supplies, and eat a lot of food.

The path hugged a long section of road stretching out of Balmaha, needing little concentration, and providing the opportunity to plan for the next stage.

Glasgow solved the bearing at least; I couldn't continue south, west was the wrong direction, so now I needed to hike east *across* Scotland. This was the plan from the start, but until Milngavie, my route on the Cape Wrath Trail and West Highland Way had stuck to a southerly bearing. I needed to turn left and eastwards.

In my usual style, I'd limited planning and preparation. Preferring an approach of a 'rough idea', I winged my adventures, which provided the chance to adapt as I went. Up to Milngavie, both the CWT and WHW had dictated my direction, but now, I had to plan each day as I travelled east. It was still unpredictable, which I thrived on, but deciding which footpaths, bridleways, and roads (the quieter ones) I intended to use had to be calculated.

This isn't as difficult as you might imagine. The UK has countless rights of way – take a look at any map and you'll see that trails cross the country everywhere. I've utilised many footpaths near to where I live over the years, so planning routes comes easily. They may not offer a direct link between locations, and weaving around is needed, but I can pick two positions on a map and link them using these paths and the odd road.

Brief, prior sweeps of the map eastward showed no shortage in rights of way. Not only was I confident of reaching the other side of Scotland, but the routing looked straightforward.

Rain fell, but I reached the Garadhban Forest just in time to fend off the worst of it. The temperature dropped alarmingly, and I stopped, scrabbling around in my pack for a long-sleeved top to wear under my jacket. By the time I reached the other side of the forest it was pouring. I put up my umbrella and quickened my pace to reach a campsite near Drymen.

The site provided a few facilities, including a hot shower and a makeshift cooking section in a barn. As I made tea and waited for the deluge to subside, the sound on the tin roof was deafening. A solitary light bulb swung in the draught, attempting to illuminate my surroundings, while water splashed outside from an overflowing gutter.

Finally, it stopped, and sunshine peeked sporadically from behind threatening clouds. I made a few trips to the field to attempt to pitch my tent, all ending in failure as the heavens opened, forcing me to retreat. Early evening, after one last confinement in the barn of misery, the sun stayed for good, and I pitched my shelter while Scotland dried and steamed around me.

A few campers dotted the field, and I chatted to my closest neighbours, who were also tackling the WHW, although travelling south to north (seemingly the preferred direction). A few of them delighted in inspecting my equipment. Most of my gear came from America, which is ahead of the UK in producing lightweight and innovative hiking equipment, so it was a novelty to them.

"What's the tent made of?" a young boy asked.

"Cuben fiber," I replied. "It's strong and light."

"I saw that movie with, erm, Bill Bryson. A Walk in…" he paused, searching for the title.

"The Woods?" I offered.

"Yeah, that one. Pretty cool that Appalachian Trail. Bet you'd like to do that one day, huh?"

"Well I have, I did it three years ago."

"Whoa! Whoa! Dad! Hey Dad! This guy's done the Appalachian Trail!" he shouted back to his tent, where his dad's head emerged after some scrabbling and rustling.

"Really?" he exclaimed. "No shit. What about the other one? You know," he sipped on his beer, "the one with Reese Witherspoon. That trail over California way. I suppose you did that as well, huh?"

"Yes," I said, smirking. "The Pacific Crest Trail. I did that five years ago, before the Appalachian Trail."

"Holy crap! Laura! Hey Laura! There's a guy over here

who's hiked both the AT and the Pacific, the Pacific, you know, the one with Reese Witherspoon in it!"

By now, his son, Duncan, was sitting cross-legged in front of me, smiling expectantly and hoping for bear and snake stories.

"Mate, have a beer!" his dad said, offering me a chilled can, while his wife Laura peered out of their tent and eyed me curiously before coming over.

I spent the next hour answering questions about my adventures, such as the usual 'how long did it take you?' 'where do you wash?' and 'how many pairs of shoes did you need?'

Dad, who introduced himself as Brian, was spellbound, as were Duncan and Laura, and, much to my delight, they were happy to exchange beer for more stories. Then he sparked the barbecue up and asked if I'd eaten.

"No, I was going to the barn to cook something but…"

"Eat with us, Fozzie – we got more than enough to go around, haven't we love?"

"Yes," Laura confirmed. "Come eat with us. You want another beer?"

Their tent was a palace. A sleeping section at the back had three compartments: one for Brian and Laura, another for Duncan, and a guest room as well. The front portion was palatial, with an inflatable sofa and chair, cooking area, and even a TV. Brian stood by the entrance, acting as bouncer to protect his property, while occasionally nipping outside to turn the meat. Laura offered me a chair, and we chatted about camping.

"I hated it at first," she explained, "didn't enjoy the dirt and lack of comforts. So, we reached an agreement."

"We did!" Brian cried from outside, muffled by the sound of sizzling steak.

She smiled. "I love walking but need my luxuries. So, I agreed to come camping more often but needed things such as seating and cooking stuff, the little luxuries so I feel at home in the evening."

She prepared a mammoth salad, bread they had picked up from a nearby bakery, and, bang on cue, Brian waltzed in with our steaks. There was an assortment of condiments, serviettes, and proper stainless cutlery.

We chatted until late, and around midnight I staggered to my tent, declining Brian's offer of a six pack to take with me the next day. They even made me a packed lunch.

The rain, as if sensing my dissatisfaction, stopped at 7am, and I packed, nursing a hangover. I longed for a nearby café where I could encourage my head back to the land of the living with coffee, bacon, and eggs, but Drymen had no amenities I could find. I was aware of the repercussions of continued evening drinking, resulting in a morning hangover, but, foolishly, I didn't care.

I dodged showers all morning, the elements constantly threatening heavy downpours. Deep black clouds sped overhead, chased by a fierce wind that didn't take no for an answer – mild one second, then smashing into my side as I turned my head and cowered, trying to stay upright. I weaved along a path riddled with puddles, attempting once more to retain dry feet.

Crossing the Gartness Bridge over the River Endrick I paused, resting my arms on the edge. The river foamed and crashed angrily, swollen by the overnight rain. The roar was deafening.

I looked back on my route and observed the landscape I

had travelled through. The Trossachs were barely visible, a faded memory cloaked in clouds and mist. Ahead lay smoother undulations, a kinder terrain rendered with more greens than greys.

Blane Water kept me company as I closed in on Milngavie. I passed Duntreath Castle, ancestral home to the Edmonstone family since 1435. The finely manicured lawns and borders seemed out of place amongst Scotland's rawness.

Some hikers passed me that morning, and I observed how fresh-faced they appeared, sporting clean clothing. I acknowledged several, their eyes shining in anticipation, excitement obvious as they spoke, and hiking with a restless, impatient eagerness to see what lay ahead.

The countryside merged into the outskirts of Milngavie, and the crunch of the trail succumbed to the silence of tarmac. Stopping at a petrol station to solve the immediate hunger crisis, my cravings centred on sugar and coffee. I grabbed an unhealthy assortment of snacks for the hotel – foods that I normally never touched, but that my body requested with urgency. As hard as I tried to concentrate, my psychological sustenance centre kept replaying clips of Danish pastries, pain au chocolat, and doughnuts. I cocked a cautious ear, wary of any subliminal sugar advertising slipped into the 80s rock playing through the speakers, but sensed nothing.

While the cashier scanned my items, I munched on my doughnut, holding it up to her as verification that, although consuming it, I owed payment.

"Any fuel, sir?" she asked, presuming I had transport despite the rucksack on my back.

"Yes please," I replied. "Black Americano, half full with half a sugar. Thanks."

She looked momentarily confused, before smiling at my joke, and placed a paper cup on the coffee machine with one hand whilst deftly operating the till with the other.

"Anything else, love?"

"I need e-fluid please. Tobacco flavour, something like that. Three if you have them."

E-fluid procured, I left contented I could vape without a blueberry in sight (or a peach cobbler), for the immediate future at least.

From empty paths I mingled with the lunch-hour crowds, my mind racing at the busyness of town. Calmness yielded to engines and chatter; the melodies of birdsong and burns diminished to barely perceptible memories.

I felt self-conscious, as I always do when I reach civilisation – thinking people were scrutinising me, judging me by my scruffy appearance and rucksack, when, of all places, at the start of the West Highland Way, they were used to it.

Douglas Street, busy with shoppers, seemed a strange place for the southern terminus of the WHW. A smooth stone obelisk attempting to blend with newsagents and coffee shops marked the end of another stage of my journey. A few hikers asked me to take photos of them, but I declined their offer to do the same for me. I wasn't happy, my mood more melancholic, not helped by the lamenting, jarring efforts of a man playing a violin on the corner.

I found the hotel and checked in, sheepishly asking if they did laundry, as I knew my clothes stank. Changing in my room, I donned my spare T-shirt and waterproof trousers and returned to hand my hiking clothes over in a bag.

"I apologise," I offered. "I've been hiking for five days; the contents don't smell great."

"It's OK, love," she replied, smiling. "I can assure you I've dealt with worse."

After eating at the pub next door, I dealt with a few necessary tasks such as charging my electronic equipment, hanging up damp gear, and leaving my shoes to air near the window. Then I showered and collapsed on the bed, falling fast asleep for three hours.

Opening my eyes, I saw rain trickling down the window outside while the wind tore at trees. I felt warm and comfortable, detached from the storm raging. Troubling sensations bubbled inside, emotions I recognised from the Cape Wrath Trail. I tried to ignore them, urging them to go away. Feelings of hopelessness and a lack of motivation persisted, worsening. I relented and stopped fighting as if foregoing an argument. I was sick of struggling, and I was becoming noticeably weaker.

A year after I returned from Scotland, and after diagnosis, I educated myself about depression and how it affects people. Signs such as irritability, sadness, feelings of inadequacy and self-hate, social withdrawal, pessimism, and loss of interest in hobbies and pastimes surface. The physical manifestations include inability to sleep, decreased appetite, lethargy, headaches, and muscle fatigue.

I experienced many of them, but the one that hit hard was a lack of motivation. The frustrating part was feeling unenthusiastic, but *knowing* that I lacked enthusiasm, and that my depression was the culprit. Despite being aware of both the emotion and the cause, the overwhelming frustration was being helpless to fight either.

I've wasted hours sitting at my desk aware I needed to write, but instead staring aimlessly out of the window until my eyes lose focus, blur, and water. The keyboard stares at me, my fingers hover over keys, primed to type words that don't come. I hope for a single grain of motivation, and, when it arrives, I hold it in my hand and watch it fall away beyond reach.

I've gone to sleep planning a walk in the sunshine the following day, even packing a bag ready. It's arranged; stop at the café first, read the paper, talk and catch up with the staff, and then drive to the South Downs, or maybe the River Arun. As I wake, and that promised sunlight creeps across my face, I open my eyes and realise the demon is back, and my plans are thwarted, for I don't have the strength to go to battle.

That demon claws at my sanity, flashes evil eyes, and mocks. It has me, it knows I can battle, but, unless I'm strong and determined, chances are I'll lose before even realising I've begun. And to many sufferers, including me, admitting defeat is the easiest choice.

It's a bully. An intimidator that approaches randomly, knowing I'm weak and unable to fight back. It pushes me around, trips and slaps me around the face, laughing. The following day it's back again and bullies harder. This can go on for days, sometimes weeks. Despite being aware of what's happening, often, as with bullying, I can't find the strength to get up off the floor, clench that fist, throw back an arm, and punch it in the face.

Then one day it's left. I don't know where – maybe it's sick, or on holiday for a week. I feel better, smile, talk to people, and my lethargy diminishes. But the bully always skulks in the shadows, biding its time.

The demon taunts using two evils. It torments – but worse than that, when it does leave, I'm in constant fear of its return.

The phone rang, startling me.

"Mr Foskett, your laundry is finished and waiting at reception."

"Thank you."

I mustered enough effort to move off the bed and sat on the edge rubbing my eyes. I collected my clothes, returned, and got dressed. There was only one thing on my mind, so I returned to the pub.

"Another beer, sir?" the barman asked, eyeing my near-empty glass.

"Yes please, and can I order the steak and chips?"

"Of course, onion rings on the side?"

"Go on then."

I'd only been in the Burnbrae Inn for an hour, and I was on my third pint. I justified my drinking on account of the dry days before, telling myself I had lost time to make up. Nagging hangover warnings were buzzing. My mobile buzzed in my pocket, and I checked the screen to see a message from Elina.

Congratulations on completing the West Highland Way! I know you're in the pub, don't drink too much and remember, be happy. x

I felt as though I were on a security camera and looked around the bar, expecting to see her hiding in a corner. Then I remembered, I'd told her when I was likely to finish and of my plans to hole up in a hotel for the night. It didn't take a genius to figure out I'd be in the pub. I typed a reply, untruthfully.

Thanks. I won't. Everything OK at home? x

Yes, a false alarm. I'm toying with coming back over to finish the hike. Doubtful if we'll be able to meet up though. Don't order any more beer. x

Yeah, I won't. Take care. x

After eating I went to the bar and ordered a double whisky, which kept my promise to Elina.

"Everything OK, sir?" the barman asked.

"Yes, good food, thanks."

"One more for the road?"

"Put another on the bill, thanks."

I weaved back to the hotel, grateful it was close to the pub. I smiled at the receptionist, struggled to look sober, and somehow made it to my room.

Accommodation in Scotland (and much of the UK) has missed out on recent television technology advancements. This phenomenon is not just apparent in cheaper bed and breakfasts, but also in mainstream places such as the Milngavie Hotel. In fact, it happens often when I stay in Scotland. It feels like they've lost a decade, as though it vanished.

The actual TV is the first giveaway. In recent years I've stayed at establishments that still sport a cathode-ray tube (for the young ones out there, that's any TV that isn't a flat screen), sat on the dressing table over a lace doily. If the proprietor has upgraded, it's the smallest, cheapest choice available. The manufacturer is unheard of – something like Sanyong, TechDisplay, Wingpong or ProTV – and shamefully displayed in small lettering near the bottom.

The controller is terrible too. It's obvious you have poor equipment when the remote offers little more than power, volume, and channel. I hate even touching those things, tentatively lifting them with two fingers to observe pizza debris and Pringle crumbs lodged in between the controls from years prior. It makes me retch.

Last, my big beef is the network availability, or lack of it.

Some B&Bs are still stuck with BBC1, BBC2, C4, and ITV.

There should be a basic right of 200 channels on TVs these days. Most of those networks are shit, but I'd prefer 200 shit stations than four. Satellite and cable have been around for years now, so you can imagine my disappointment when my plush, none-too-cheap hotel offered just 10.

I understand the investment issues in upgrading, especially with technology moving quickly. With the larger chain hotels offering 100, 200, or more rooms, that's a hefty outlay. If relenting to customer demands of a 50-inch-wide-screen with the latest definition, at £2,000 each, that's an expensive purchase.

Couldn't they at least warn me in reception first?

> *Dear guest,*
>
> *We apologise as our TV sets are ancient. Terrible really, we should upgrade, but it's pricy; we have 156 rooms you know, that's a lot of money.*
>
> *Just to inform you, we only have four channels, and one of those is local Scottish news.*
>
> *We haven't cleaned the remote controls since we last upgraded 13 years ago.*
>
> *The TVs are Wingpong though…*

I think I may start a new organisation – *The UK Accommodation Television Upgrade Lobby*. That's UKATUL for short. It's catchy and sounds like a TV manufacturer. I'd need a catchphrase, a selling point.

Demanding, lobbying, and supporting your fundamental rights to a fair and modern accommodation viewing experience.

I took one last look at the screen, with the local weather girl filming on location in Glasgow, and, despite the sound being on mute, her umbrella told me everything.

In their defence, the hotel had laid on a fantastic breakfast. It was a hiker's dream. First the teasers: muesli, granola, cereal, and fruit. Endless coffee and tea, and all-you-can eat hot stuff. Bacon, sausage, hash browns, mushrooms, tomato, eggs how you like them, black pudding, and toast. Some local delicacies were displayed, and, while munching on granola, I browsed my mobile, researching the Scottish breakfast.

First up was Lorne sausage, also known as square or sliced sausage and made with ground pork, beef, rusk, and spices. Compressed into a rectangular tin to make a meat-shaped loaf, it's then sliced and fried. Imagine a block of Spam, and you get the picture.

Opinions vary as to its name. Some say the region of Lorne, in Argyll, is responsible, while others swear it's named after the comedian Tommy Lorne. Alex D. Cochrane, the author of the Adcochrane website, in his 'Ten glorious and dubious Glaswegian food delicacies', describes it as 'pure crap'.

A potato scone, also called a tattie or tottie scone, is made with mashed potato, butter, flour, and seasoning. The mix, rolled to a roughly 7mm-thick circle, is then cut into four segments like a slice of pie and cooked in an oven or hot pan. I've had it many times – the combination of potato and flour is delicious.

Now as for the Arbroath Smokie, even the name sounds appealing. This is a haddock fillet, salted and left overnight, then hung in sealed barrels over a hardwood fire for 30 to 60 minutes. Auchmithie Harbour is considered the true home of the Smokie, and the name itself is safeguarded under the

EU's Protected Food Name Scheme.

Finally, a buttery, also known as a rowie, or rollie, is a flaky, savoury roll like a croissant. Created in the 1880s for the fishing industry, its high fat content due to the large quantity of butter meant it kept for longer at sea.

After the distance I'd walked along the wilds of the Cape Wrath Trail and the West Highland Way, I was looking forward to the relative luxury of passing regular towns and villages. I had no predetermined course to follow, and the chance to make the route up as I went was exciting.

Each approach has its own advantages; a marked trail means a hiker can relax, safe knowing that, as long as they pay a modicum of attention and observe signs, they can't get lost. However, as much as I love existing routes, I revel in mixing things up and inventing my own, personal passage.

I checked the possible route options on my phone while succumbing to one last coffee, sitting outside a café. I had two choices: either plan the entire section between Milngavie and a town called Kirk Yetholm, my target on the other side of Scotland, or make it up as I went. Kirk Yetholm was around 150 miles distant, depending on which paths I took to get there, and a mile from the border with England. I chose this destination simply because I'd heard it was a nice place, was due east, and had a youth hostel. It also linked up with the Pennine Way, stretching south into England, another trail I was deliberating.

Planning the entire route had two problems. First, it would take me hours, especially with failing eyesight on a smartphone. Second, it clashed with my desire to get random; I wanted to adapt and change as I went.

I settled on the latter option, and each evening I took a few minutes to scan ahead on the map, picking footpaths, bridleways, or quiet roads according to what I needed. If I wanted food, a shower, or laundry facilities, I could adapt and head for town. Equally, if I desired some peace, I could stay away from urban areas. The whole process, tweakable as I progressed, was perfect.

Starting at Milngavie, I took a swig of coffee and squinted into my phone. My eyes focused on a thin, blue strip leaving town to the east, and I became excited. The Forth & Clyde Canal meandered all the way to Edinburgh. Now, the big city lay too far north, but the waterway solved the initial direction choice. To get there, I found the Kelvin Walkway, a footpath that connected Milngavie to central Glasgow.

The River Kelvin flowed over and said hello, and I reached a bridge under the A879. The river and walkway carried on towards Glasgow, but the road signalled my turn. I crossed a wooden bridge and carried on by the road side for a few minutes, then took a left onto the quieter Balmuildy Road.

For two miles, the Antonine Wall, built by the Romans around AD 140, was visible to the north. The remains stretch 37 miles from Old Kilpatrick on the west coast to Bo'ness in the east. It was once the northernmost frontier of the Roman Empire. Unlike the well-known Hadrian's Wall, constructed of stone, the Antonine Wall is a turf structure, fronted by a ditch with a dirt track behind to aid the passage of troops and transport. The entire length is dotted with the remains of forts.

Balmuildy Road, despite being narrow, had its fair share of traffic, and I spent much of the time wedged against hedges as vehicles approached from both directions. It was annoying, but a dirt track appeared by a bridge after three miles, and I emerged onto the Forth & Clyde Canal.

The transformation was immediate as the traffic noise faded and I alighted into more discreet surroundings. The water was motionless save a breeze misting the surface, and my passage alternated between straight sections and gentle bends. Either side stood a low hedge, and occasional trees funnelled me onwards.

Water is incredibly calming, and it didn't take long to have the same effect on me. Town and the roads forgotten, I felt at ease, not only back in the greener lands, but by the water.

I'm fascinated by canals and narrowboats, and I dream of living afloat one day. I was in my element, not only relaxed but eager to travel along the waterways for as long as possible and experience the culture that has grown up around them. A narrowboat appeared ahead of me. Passing, the owner, resting one hand on the tiller, raised his cup of tea with the other.

"Morning!" he cried. "A good one too!"

"It is!" I called back.

The Forth & Clyde Canal opened in 1790, and, as with many canals, transported both goods and sometimes passengers. At 35 miles long, it runs between the River Carron at Grangemouth to the River Clyde at Bowling. The advent of the railways, which shipped merchandise quicker and in larger quantities, signalled the beginning of the canal's end. It closed in 1963.

In 2000, National Lottery funds regenerated both the Forth & Clyde and the connecting Union Canal, once more providing a continual course between both ends. The Forth & Clyde pathway runs alongside, which I planned to use for the next stage of my trip. I was to follow it to Falkirk and the impressive Falkirk Wheel, which lifts boats between the Forth & Clyde to the Union Canal. I intended to keep to the water until then, which made navigation easy and suited my wish for as little planning as possible.

There were other positive sides to canal walking. Bridges meant plenty of opportunities for shelter, ideal as I played cat and mouse with the rain all day. I'd never experienced so many showers, not knowing when one would end and another begin. I used the bridges to plan each stage, studying the map to find the next dry spot.

My route was also flat! After the extremes of the Cape Wrath Trail and the West Highland Way, it was an absolute joy to rest easy knowing I had no inclines. My body approved of this newfound freedom, and I looked forward to walking further each day with less fatigue.

Access to amenities was frequent. Instead of calculating days until my next resupply, it was now only hours. Villages and towns dotted the length of the canal, so I had regular food stops, either to buy in stores or eat in cafés. My pack weight plummeted, and I enjoyed a lighter load.

I had constant drinking water. Canals are not ideal sources, but, with the aid of my filter, I didn't need to carry as much. I realise canal water will never be up there with Perrier, but it tasted fine – albeit with muddy notes and slight undertones of diesel.

Finally, the history promised great things. With the Forth & Clyde being well over 200 years old, many of the fringe buildings and associated infrastructure proved interesting.

Despite sheltering from the intermittent showers, my path of flatness delivered easy walking, and a familiar pattern emerged. The countryside succumbed to villages and towns along the canal's course, and I saw an increase in the number of dog walkers the closer I got. In fact, it acted as an early warning signal that a town was approaching. From the serene green in between, dogs ran alongside the canal with their owners, who clutched a mix of rubber balls, poo bags, and leashes.

The Campsie Fells and Kilsyth Hills folded up to my left as my shadow lengthened towards evening. The sun behind me from late afternoon offered a good sign of the time, and that, coupled with tiring muscles, signalled the start of searching for a place to spend the night.

I had noticed a surprising lack of suitable pitches that day. This is not something I'd expected, reasoning that the towpath should be flat. What I hadn't bargained for were the hedges and trees which encroached on or near the trail, leaving little room to squeeze my tent. On the other side of the hedges were fields, green with corn and offering no options. I eventually squeezed into the tightest of spots against a hawthorn bush, with uneven ground, my sleeping mat tempering the lumps and bumps.

Being away from town, the evening was quiet. A few ducks splashed. As I looked to my right, the sun was sinking above the west end of the canal. The glow reflected in the water, turning the Forth & Clyde into a strip of bright orange, brushed with strokes of dark blue, and black from a few clouds. It was magical, motionless and blissfully silent.

Chapter 11

Dutch Barns, Pedalos, and Abandoned Houses

I watched a grasshopper just a few inches from my face, nothing separating us except the tent's insect mesh. It climbed up a plant stem, stopping to stretch its legs while eyeing me, then jumped and disappeared into the grass.

A breeze barely murmured while the canal mirrored, motionless, imitating the sky flawlessly. A perfect day to be by the water, I waited for a disturbance on the surface to confirm I wasn't dreaming. My hope was met as two swans chaperoned their cygnets, gliding effortlessly past, tiny ice skaters on the silver. Plants broke through the water, searching for light, and swayed.

Approaching a bend, I heard a strange, rasping sound. It continued until I rounded the turn and a moored boat appeared. A young couple looked up.

"Morning!" the woman cried, and the man waved.

"Hello! Spot of painting preparation?" I asked.

"Yup," he said, taking a swig from his mug.

Their boat was impressive, albeit in need of some repair and attention.

As I paused on the bank, she continued. "We only got her yesterday, it's our first day on the water and we're going to live aboard. It's super exciting! But, I realise how much work needs doing."

The bloke wiped his hands on his shorts and turned to her.

"More tea?"

"Oh God, you have no idea how thirsty I am," she said.

"Can I offer you a drink?" he said to me.

"Er, well, yeah. Why not? That'd be great, thanks."

I sat on the bank, and we introduced ourselves.

"I'm Mary, and this is my husband Steve."

I shook hands with them both. "Fozzie, nice to meet you."

"I don't suppose you know anything about painting?" Mary added.

I smiled. "A little, yes." She handed me a tea, and I took a long, grateful slurp. "Ah, great tea. I used to be a decorator."

"Jackpot!" she cried. "Steve! He's a painter!"

Steve's head poked out of the rear hatch, his face beaming.

"We have questions! Many questions!" he called.

"I'll tell you everything you need to know in exchange for a guided tour. I've always wanted to see inside a narrowboat."

"Deal!" they both cried in unison. "Come aboard!"

I hopped on, and they showed me their new home. We started at the bow, in the living room, or saloon as it's known. Then the kitchen, on to a corridor with a door on one side leading to the bathroom, and finally a bedroom. A hatch at the stern led to a deck with the tiller and other controls.

"We're going to cook breakfast," Steve announced, looking my way for a reaction, to which I did my best to look pleased. "Would you like to join us?"

"I'd love to, thanks."

As the aroma of bacon and eggs drifted from the window, I answered as many questions as possible. Their first problem was sanding the entire boat to prepare for new paint. I pointed out that preparation is everything, and that they should take the time to do it right.

Eventually, he handed me a plate with bacon, eggs, mushrooms, and hash browns. Mary had set up a table on the tow path with chairs and a huge pot of coffee, and as I sat she squeezed orange juice.

"This is fantastic. I'm glad I said hello earlier."

"We appreciate your help," she replied, handing me a juice and enquiring how I liked my coffee.

"Black is fine thanks, half a sugar."

"Where have you walked from?" Steve enquired.

"Cape Wrath," I said, chomping on bacon with one hand and stirring my coffee with the other. "This is tasty by the—"

Mary cut me off. "Cape Wrath is on the north-west coast! That's miles away!"

I grabbed my diary, poking from a side pocket on my pack, and looked at it.

"444.5 to be exact," I said. "Give or take."

"Shit! Why? Where you heading?"

"Kirk Yetholm, but I may carry on further, depends how I feel. The actual plan is to cross Scotland – I always wanted to do it. Why not? I love hiking, it's an independence thing.

I enjoy the solitude, the escape, lack of schedules and agendas. You know, the freedom. Exactly what you two are doing, only I'm walking and you're boating."

"We also get regular bacon access," Steve commented, which was a fair point.

"You do, and damn fine bacon it is!"

And there we stayed for an hour, a unique start to a day's hiking: eating a cooked breakfast on a table, next to a narrowboat by the canal.

"We're renegades, that's what we are," Mary chipped in.

"I like that word," I replied. "It's a good fit."

She disappeared into the boat and emerged with a dictionary.

"Renegade," she began, "a person who abandons the religious, political, or philosophical beliefs that they used to have and accepts opposing or different beliefs. They missed out accepting a different *life*, but it works."

"It does," Steve said, refilling my coffee. "We are the renegades!" And we raised our mugs and laughed.

"To the renegades!"

I left them, reluctantly, and ventured east. I daydreamed of owning my own narrowboat called *Contentment*, cruising the waterways, constantly on the move to indulge my nomadic tendencies.

I heard the suburbs of Falkirk muttering over the hedge that separated us. Cars, the clanging of industrial estates, a couple arguing. Detached, I was in another world by the water.

An hour after leaving Steve and Mary, an iron beast rose high. The Falkirk Wheel, one of the strangest engineering feats I'd ever seen, stood an impressive guard.

As part of the regeneration of the Scottish canal network, the Falkirk Wheel was constructed as a missing link between the Forth & Clyde and Union canals. The height difference between the two is 115 feet. A pair of locks on the upper Union Canal make up 36 feet, leaving the Falkirk Wheel to deal with the remaining 79 feet.

It does this by a rotating cradle design, and the best visual comparison I can offer is a Ferris wheel. Instead of many passenger cars, there are just two, but they're huge, as you can imagine they'd have to be for carrying a narrowboat. In fact, each one can hold 250,000 litres of water, weighing a whopping 500 tonnes, and is capable of carrying up to four 66-foot boats.

Descending from the Union Canal to the Forth & Clyde, boats enter the gondola, and the watertight doors close. The wheel rotates 180 degrees until the gondola reaches the canal below, where a dry dock floods to allow onward transit. Going up, the process is reversed.

I stood for 30 minutes observing, watching this huge steel monster rotate. It made little noise save the engine's gentle hum and executed its work gracefully. It surprised me that such an immense construction was capable of such fluidity. Inspirations for the design included a whale's ribcage, a ship's propeller, and a double-headed Celtic axe. The head architect, Tony Kettle, described the wheel as 'a beautiful, organic flowing thing, like the spine of a fish'.

Realising my time by the water was drawing to an end, I continued along the Union Canal. It steered a north-east course for Edinburgh, and, apart from my dislike of cities, even one as renowned as the Scottish capital, my direction begged a more south-easterly tangent towards Kirk Yetholm.

Sitting on a bench, I watched two swans and seven

cygnets glide past effortlessly, trying to catch a couple of canoeists. The route needed assessing, so I studied my phone while munching on all I had left in my food bag: stale bread.

It had been a while since my dystopian interludes. The Cape Wrath Trail and West Highland Way were memories now. Even my frustrations back then with the lack of trail made me chuckle. Those paths were gone; my route now was luxurious in hiking terms. The canal towpaths were excellent, and in between I mixed it up on other footpaths, bridleways and quiet, minor country roads. My plan of evaluating the route daily was working well, but now the canals were ending I needed to figure out a course to Kirk Yetholm.

The immediate issue was avoiding large towns, of which Livingston was the first. It would prove difficult to circumnavigate, no matter how much head scratching, so I plotted a route through the outskirts and hoped they'd be pleasant enough. I dreamt of a hot takeaway that evening, especially as my food message centre kept suggesting chips with curry sauce.

So far, I loved the Scottish minor roads. Fewer people meant fewer vehicles. In fact, it felt unusual to hear a car engine on those roads – they were blissfully quiet.

Arriving at the outskirts of Linlithgow, I found the road heading towards Riccarton. I passed through West Binny, Bankhead, Craig Binning, and Dechmont before reaching the noise of the M8. A tunnel led me under the motorway where I arrived in East Calder.

I was tired, and, thanks to my pause while talking to Mary and Steve that morning, it was already 8pm. I waited in the fish and chip shop while debating my sleeping options that night. Opportunities in town were going to be limited, so I headed back to the countryside, satisfied after my chips and curry sauce.

After a mile I struck accommodation gold. A Dutch barn, half full with hay bales, sat in the corner of a field. Dutch barns may not seem ideal places to bed down, but they offer benefits: warmth, security, and shelter. Hay is a great insulator, and, even though it was a warm night, it reminded me of colder times spent in barns over winter. It doesn't feel too cold with a few hay bales stacked around you, and they also keep you from prying eyes. I'm sure there are legalities when it comes to barn camping, so staying out of sight is a wise idea – but on the occasions I have been caught sleeping in one, the farmer has, at worst, wagged his finger disapprovingly, but smiled and let me be.

I've spent the night in many strange places during my travels. During my thru-hike of the Appalachian Trail I discovered the delights of sleeping under freeway bridges. Most entailed scampering up a steep slope to reach a level platform a few feet wide. I always stayed completely dry, and the road noise from above was surprisingly unobtrusive. The cops had no idea either.

On a cycle trip through Italy with my mate Jeremy (who'd accompanied me on the Scottish winter course), we stopped by a beach to look for a suitable overnight shelter. The local funfair, closed for the season, offered a solution as we slid under a fence to a compound where several pedalos were stacked. Three were accessible, piled on top of each other, so I took the bottom as Jeremy climbed up to settle in the one above. I soon discovered pedalos weren't ideal resting areas; every time he adjusted position, a shower of dried salt dislodged and covered me. I moved and spent an uncomfortable night hunched between the fence and a fortune telling machine.

On that same cycling trip, just before crossing the Alps, we arrived at a Swiss village called Guttannen. I'd read that if you're stuck for a place to stay, the local church could be an option. We knocked on the nearest door and, in poor German, asked where we could find the local priest. As if by a miracle, it was the priest's house, and he was more than happy to let us into the church over the road. Unfortunately, it proved another bad choice, as the bells tolled every hour, the first of which so startled me I shot bolt upright and cracked my head on the pew I was lying beneath.

I don't frequent abandoned houses any more, especially after an incident on El Camino. I'd had a long day on the Aubrac Plateau in France, and sleep options looked limited. Unable to pitch my tent because I'd damaged it the night before, a flickering light caught my eye in a clump of woods. As I approached, to my surprise and delight, I saw the outline of an old building a quarter of a mile away, with an illuminated window.

When I arrived, the light had mysteriously vanished. Now I'm not in the habit of exploring empty, dark houses in the middle of nowhere, especially with flickering lights. Let's face it, it's the opening scene to a horror movie. But I assumed the light indicated other hikers who had thought the same as me and investigated, looking for somewhere to shelter.

The moon illuminated the house weakly, casting shadows on the old stone walls, as ivy crept everywhere. Despite my apprehension about the light, thinking someone may already have been inside, I did a quick scout of the two rooms downstairs and three upstairs. It was empty, so I settled in what appeared to have been the kitchen, cooked a meal on my stove, and fell asleep.

An hour later I woke, frightened by the kitchen door slamming shut, opening again, then banging shut again in a wind that had appeared from nowhere. I wedged a piece of wood under it and went back to sleep. Startled once more by the banging of pipework from upstairs, I woke again. By now, as you'd imagine, I had the heebie jeebies. I hesitantly crept to the other room where the clanging was coming from, and, as I entered, the sound stopped. That was enough. I scooped up everything I owned and fled through the woods to the relative safety of a field. Luckily, it wasn't raining, and I spent the rest of the night sleeping in the grass. As I said, I don't do abandoned houses any more.

Tenting is great, but occasionally I prefer alternative accommodation. Hoboing is the American expression meaning to wander from place to place with no permanent home, sleeping wherever possible. You'd imagine such a lifestyle unattractive, but it's a wonderful way to live (although maybe not in winter!).

I propped my back against a hay bale, plugged my earphones in, and selected one of my favourite albums. After warming up the piano, Roger Hodgson's voice joined in. As I listened, the lyrics sounded appropriate.

> *When I was young, it seemed that life was so wonderful, a miracle, oh it was beautiful, magical. And all the birds in the trees, well they'd be singing so happily, oh joyfully, oh playfully, watching me.*
>
> *But then they sent me away to teach me how to be sensible, logical, oh responsible, practical. And then they showed me a world where I could be so dependable, oh clinical, oh intellectual, cynical.*
>
> *There are times when all the world's asleep, the questions run too deep, for such a simple man.*
>
> *Won't you please, please tell me what we've learned, I know it sounds absurd, but please tell me who I am.*

I stopped the music and lay there thinking about those lyrics for an hour. Life was wonderful when I was young too, before the stark complications of adulthood arrived – I was also carefree and happy. I listened to the birds in the trees when I escaped to my local woods.

Then it went wrong, and they sent me away as well, to school. I wasn't free any more but was expected to attend this sombre building every day, listening to adults tell me about the world and how to behave.

I was scared. I didn't like the routine, hated being with others, and resented having to learn subjects in which I had no interest. But most of all, I despised having restrictions imposed on me and being powerless to protest. The overwhelming pressure of compulsory obligation angered me, and it still does today. I hate authority.

Feeling trapped, I realised that this was just my first school, and that afterwards I had junior school, then comprehensive as well. Suddenly, my life was a locked cell. I knew my parents worked, but never thought I'd have to do the same. Swiftly, that prison sentence began to look very scary.

I think that's why I ended up how I did. The enforced schooling, training, and jobs I dealt with begrudgingly for years moulded me into someone who resented authority. Of course, I was too young to understand at the time. Now my hatred of the system we live in – the politics, the rules, regulations I'm expected to abide by – angers me. I'm essentially a law-abiding person. I understand that not stealing, avoiding fights, and driving on the left-hand side of the road are inherent to a decent society. But, my anger towards the government and their irrelevant policies, when there are far more important issues to consider, scares me.

Now my frustrations were manifesting in rage. I was

never a person to lose control of my emotions. I didn't know what was happening, but Scotland, where I had come to escape my moods, was making matters worse. The outdoors, my last failsafe, the place that always made everything better, was failing. I felt helpless – if the outdoors couldn't help, what else was there?

I wanted to be back in the woods, listening to the birds.

At 3am I bemoaned my choice of shelter. The sky lit up, and as I peered out, lightning forked over Scotland. Thunder crashed, not gentle rumbles but fierce cracks that made my hair stand upright (more from being petrified than the static electricity). Suddenly, a metal barn didn't seem the best choice for shelter. The deafening cacophony on the roof made me quit any hope of getting to sleep.

By morning it had stopped, but angry skies loomed and cast menacing eyes. I plodded through mud and puddles to a railway line where my map had marked a tunnel that didn't exist in reality. I pondered climbing the fence, but it was too high, especially with a rucksack, so I retreated.

The rain started again, chased by wild gusts that knocked me around like a bowling pin. As I pulled a bag of granola from my pack, the contents emptied through a hole onto the ground. At that point, I knew it was going to be a bad day.

My phone vibrated. It was a text from Elina.

Hey. How is it going? she said.

Honestly? I replied. *Fucking shit. It's days like this I wish I'd never left!*

What's wrong? You want to talk?

No, it's fine. I'll be OK. Just an off day that's all.

She didn't respond, and I felt guilty for being so curt.

I hadn't fallen in the pit for a few days, because as the weather had improved so had my outlook, but now I was shaky. A few negative events, such as spilling food, reading the map incorrectly, and being wet, were all it took for the situation to get on top of me. I'd been through worse and always managed to come out smiling, staying calm and rational. Now I was annoyed and back to the habitual feelings I'd experienced during the dark days on the Cape Wrath Trail.

I detoured from the railway to Kirknewton, where I restocked with food. Picking up a B road, I walked for 30 minutes before coming to a junction with the A70. I had two further miles along the road before a trail took me over the Pentland Hills via a track called the Thieves Road, which climbed to a modest 1,300 feet.

The A70 had no path by the side so I played cat and mouse with the traffic, mainly a lot of big trucks. With nowhere to shelter except the occasional tree, I tried to judge my progress and reach the next one before I got soaked from the vehicle spray. I failed, and my waterproofs became saturated. Wiping a mix of water and filth from my face, I spat out something that tasted of diesel.

I was livid, and I couldn't believe how angry I'd become. I made the turn-off for the Thieves Road, stuck my middle finger up at the traffic, then sat by a wall kicking stones and cursing. All I wanted was a cigarette and some beer.

My phone buzzed; it was Brian, a workmate at home.

Fozzie, got a load of decorating work starting next week! Need you here!

I read it and sighed.

You know I'm in Scotland. Won't be back next week. I told you I'm not available.

I replaced the phone in my pocket, thinking that would

be the end of it, but it buzzed again.

Yeah, but it's good work! I'll put you down as a yes then, you know it makes sense.

Seething, I typed another reply: *What part of 'I won't be back next week' is it you don't fucking understand exactly Brian? Leave me alone. Seriously, you're pissing me off.*

I hurled my phone against the wall and watched, sickened, as it smashed into several pieces. I collected the fragments but, realising it was irreparable, left them in a nearby rubbish bin after rescuing the SIM card. Releasing my frustration hadn't made me feel better, now I felt worse – and I needed a new phone. I retrieved my spare, turned it on, and hoped all was well after not using it for months. The home screen appeared, and I opened the map to study the route.

My gaze rose, focusing on the Pentland Hills as clouds raced over, tickling the summits. I had seven miles until the village of West Linton, where I knew there was a café. I made plans to crack it out in two hours and be sipping a strong coffee in the warm.

I followed the fence over spongy ground, dotted with planks bridging black, boggy mud. The path initially cut a straight line for a mile before bisecting East Cairn Hill and West Cairn Hill.

The Thieves Road was quiet, which belied its past. As the name suggests, it was a haunt for bandits who stole not money but a different commodity: cattle.

For 300 years, up until the early 20th century, the drove road was one of the busiest in Scotland. Thousands of cows every year were herded to the English markets in exchange for sheep and money. Now it was quiet, the noise and bustle silenced, the thieves gone.

The summit is called the Cauldstane Slap. It is thought

that Cauldstane refers to the strong winds that whip over the pass, and indeed Slap does mean a pass.

Slapped was suitable, and exactly how I felt at the top. Gales gusted randomly, making it difficult to judge. I braced myself for winds that didn't arrive and got tossed around by those that did.

I could see West Linton on the map (and almost smell the coffee) and chuckled as I noticed the wonderful names of the hills; Muckle Knock, Grain Heads, and King's Seat. It was early afternoon as I arrived at the charming village, wandering through streets lined with stone buildings. Nothing much stirred, my feet the only source of sound. I found The Olde Toll Tea House, pulled the iron catch, and entered. Leaving my pack in the lobby, I took a chair in the small, cosy room.

"Weather not too good for you hikers at the moment!" came a cry from the kitchen.

"No, it's not," I replied. The elements were always a decent conversation starter, so I continued. "Have you seen the forecast?"

A middle-aged man sporting coiffured ginger hair, not unlike Donald Trump, appeared holding a menu. A tea towel draped waiter-style over his arm.

"Aye, it's looking crap," he said, apologetically. "I'd say you're going to get a wee bit wet." He left the menu. Craving cheese, I spotted the Welsh rarebit hiding, for reasons unknown, in between the cream scones and tea selection.

"Strange location to place the Welsh rarebit?" I said. "But it sounds good, and a black Americano please."

"Printing error!" he cried amidst the clanging of pots in the kitchen. "I pointed it out to the printers, but they denied liability. Bastards!"

A steady stream of locals arrived and left, ordering their

daily coffees, and I chatted to a couple of Dutch cyclists who looked how I felt: bedraggled.

I cowered like a scared kid as I opened the door to leave, expecting the rain to pelt me once more, but, despite threatening clouds, it had stopped. I continued along the drove road, which shared its course with the B7059 before crossing the A701. Negotiating between Drum Maw and Hag Law, I eyed the cloud base on both summits, but before I had a chance to gauge the weather it started raining again. I increased speed, seeing a large forest just after Green Knowe. Tall pines offered shelter, shielding me from the storm. The temperature dropped, and I added an extra layer, shivering. Trees blocked the light as I carried on through dull and dingy surroundings.

Peebles, the next town, was four miles away, but there was a stretch through open ground where I was sure to get soaked, so I opted to camp in the forest just by Kilrubie Hill. This proved a wise choice – Scotland was about to get hammered by another storm.

The trees were so close together that finding enough space to pitch my tent was impossible, so I settled on the edge of a track. Tired and unable to keep warm, I got into my sleeping bag and fell asleep. Once more, in the early hours, I woke up startled by thunder and lightning. The wind worried me, constantly tearing at the guy lines as the fabric flapped loudly.

Then my shelter split.

A gaping hole tore open. As rain blasted in, I scrambled for my head torch and threw my gear to one end. I took down my trekking pole, which acted as an internal support, so the tent lay flat, and attempted to wrap the fabric around me.

There I stayed, awake and scared for six hours. Terrified

of a lightning strike – Kilrubie Hill was up at 1,300 feet and the trees didn't help – I clung onto the tent while peering out of the mesh. Rain lashed the ground, which had transformed into a torrent as water gushed past. The wind roared furiously, tearing at the pines, which swayed erratically. I remained dry and warm until 7am when the storm relented. And as if by apology, the sun broke through as I got up.

Shoppers eyed me warily, like they would a caged animal, as I walked through Peebles. Dirt covered me where I'd slipped and fallen in a muddy field. The smell I gave off suggested there may have been cow dung in that mud as well. I must have looked grumpy as I squelched through the streets. One mother, looking alarmed, guided her children away from me.

I called a mate who was looking after my gear.

"I wrecked the tent. Can you send the spare up to me please? I need it DHL or similar, preferably tomorrow. Is that OK?" I said.

"Of course. What happened?"

"Got beaten up by a storm, it tore the tent and it's useless. Did I leave the Contrail with you?"

I heard rustling. "Yeah, Contrail is here. Will that do it? Anything else?"

"Yup, that should do it. No, nothing else. Send it to the Cross Keys Hotel in Peebles please, mate. You're a champ, thanks."

I hung up and made my way to the hotel. I left my boots outside and did my best to make myself presentable.

"You look like you need a rest, sir!" the receptionist exclaimed. "Just the one night?"

"Probably," I said. "Maybe two, depends how my plans go. Do you do laundry by any chance?"

"The machine is being repaired I'm afraid, sir. But Northgate Laundrette is very close." She handed me a map.

"Thanks, I'll find it and let you know in the morning if I need another night. Is that OK?"

"That's fine, sir. You're in room seven, first floor. Have a nice stay."

The room was excellent, but it couldn't lift my mood. After a laundry trip, I showered and fell asleep. I woke feeling dejected, a feeling I had come to associate with the pit. One of the first warning signals was needing sleep, and I didn't want to get up. Bed was a safe place, somewhere to hide, although I knew it wouldn't solve my problems.

My mind returned to the woods and the birds I'd loved as a child. I heard them singing, remembering the sunlight filtering through the trees. I evoked that happiness and realised joy had been a constant companion during my childhood. Save the occasional argument, becoming sorrowful after watching a sad movie or other such situations, those cheery feelings didn't leave me. The gloomy moments never stayed for longer periods; they didn't keep returning, poking or mocking me.

Depression ridicules. It knows it can get away with it because of its strength and our weaknesses. Even if I could lash out, what could I lash out at? It may be a bully, but it's not a physical tormentor. I couldn't reason with an invisible oppressor. It buries itself in the crevices of the mind, the unreachable places – the constant intimidator.

My thoughts racing, I sat on the edge of the bed, unable to move, and then started to cry. I was psychologically exhausted. Everything I needed to do to reach England seemed insurmountable, and I had nothing left to contest that challenge. One side was continuing to fight, but the other had surrendered.

Elina returned and spoke softly.

Depression isn't voluntary, Fozzie, we have no choice.

If I did have this mental illness, why did I have it? When had it started? What had happened in my life between listening to birds and sitting in that hotel room? My mind raced back, trying to remember similar feelings in the past. I recalled emotional events on the Pacific Crest Trail, the Appalachian Trail, and other incidents at work or with friends. No memories of my fragile state were obvious before the Pacific Crest, seven years ago. My life before then, albeit with its highs and lows, had moved along happily.

My thru-hike of the PCT had been an amazing experience, but it was fraught with periods of little motivation. I spent too long in towns drinking because I had no impetus to get back on the trail. I nearly failed because I was so late. In fact, I had a three-day stay in a town called Packwood to sit out the rain, unable to muster the enthusiasm to leave the hotel, and I even referred to that period as the 'Packwood depression'. I'd used the word *depression* seven years ago! Had I chosen it as an analogy, or had I unknowingly realised?

I picked up a magazine from the bedside table and flicked through it. Pausing on one page because of an interesting photo, I noticed a quote highlighted in the middle of the text.

Sometimes even to live is an act of courage
Lucius Annaeus Seneca

Kirk Yetholm was 44 miles away. The border with England and the end of my hike was another two miles from there.

I refused to be in Scotland any more, felt tired of hiking, and didn't want to carry on south.

I just wanted to sleep. I wanted to hide.

Chapter 12

St Cuthbert and the Castrojeriz Herb Parcel

My plan was backfiring. I had come to Scotland to escape, an exodus from discontentment. My happier times centred around hiking trips: El Camino, the Pacific Crest Trail, the Appalachian Trail, and other adventures in the UK and Europe. I reflect on them with pride, grateful for those periods in my life when my spirits soared.

They were my getaways from life's mundanity. I had longed to reach those places, revelled in the time I spent there, and cherish happy memories. They were my treatment, my failsafe.

But like becoming tolerant to overused medication, my countryside therapy was weakening. This, in turn, made me angrier. I'd been unhappy for a couple of years and didn't know why.

I was sick, and my outdoor failsafe was fading.

Peebles wasn't helping either. The town was quaint but solemn, hiding under a low mist, punctuated by occasional drizzle. July felt more like October. People emerged from shops huddled under umbrellas, pulling summer clothes tightly around them, dashing to their destinations. The streets glistened, each cobble shining, a million staring eyes scrutinising me.

I'd booked into the hotel for another night. My lethargy and the foul weather justified the decision to stay put and wait for my tent. At least now I could be miserable and hide in my room.

When reading a newspaper or magazine article, I work my way through but often hurry over the last paragraphs, or I give up altogether. I don't know why, and it annoys me. Sometimes this laziness carries over to other areas. The gym is a classic example. I'll do an intense hour on the cross-trainer and become frustrated towards the end, losing interest. I'm aware of the feeling, and I persevere through it, but I wonder why it happens.

My mind was playing a similar, dangerous game with the hike. Nearing the end, my enthusiasm was waning. Despite my eagerness to continue on to the Pennine Way into England, my psyche seemed to have reached a decision, trying to persuade me to finish sooner. The newspaper article was over, the cross-trainer completed, or so my head thought.

I holed up in the local café for most of the morning, flicking through the paper and ordering coffees to warrant my seat. Moisture misting the windows allowed me to hide, a distorted barrier against the world outside. I encouraged drops to run down the glass, creating patterns. A bell dangled above the door, chiming when anyone entered and left, as a stream of locals collected drinks and pastries.

My phone rang, and I checked the screen: MUM.

"Hi Mum."

"Hi Herb! Where are you?" I should point out that my parents call me 'Herb'. It's short for 'Herbert', as in a 'right little Herbert'. This is a child always up to mischief, although I deny all accusations.

"A place called Peebles. I'm holed up in a café, taking a day off."

"Is it stormy?" she said. "We saw the weather report last night and Scotland looked wet."

"Yes, it's shit," I replied honestly.

"Are you OK? You sound a bit miserable."

"Yeah, sort of. I hate it when I'm hiking and it's raining. I'll be fine when the sun comes out."

"Well, you know how you are sometimes." This was my mum's favourite phrase, and she used it often; but as she continued, it made more sense. "You do have these periods when you're unhappy, it worries me. You're up and down a lot but you're walking, that's where you like to be."

We chatted further before hanging up, and her words lingered.

You do have these periods when you're unhappy.

I kept missing the clues, and it wasn't until months after Scotland that I realised. The hints, prods – not just from my mum, but Elina, and others as well. The signs were everywhere.

Imagine this. I'm walking along a street, and everyone is stationary, motionless, looking my way. They point and wave to get my attention, and they each say something.

"Hey Fozzie! You look unhappy."

"Look at me, mate, give me eye contact, I'm trying to help you."

"You're feeling low for a reason, Keith."

"Can't you notice a pattern here, chap?"

But I'm wearing horse blinkers, so I can't see anyone. I'm unable to hear because my headphones are playing music, and my clothing is thick, so I don't feel them touching. I walk the entire length of the street oblivious, turn the corner, and disappear down an alley where no-one ventures.

A few months later, when I did realise I had depression, I couldn't help but smile at the clues littered behind me.

"Do you mind if I sit here? Excuse me? Sir?"

I snapped out of a deep daydream and looked at the woman, realising she was talking to me.

"Sorry, yes. I mean no, miles away. Help yourself."

I smiled and turned to the window as sunlight flooded the café. Wiping a circle on the misted glass, I peered out with one eye. The clouds had cleared, and, as Peebles started its drying cycle, people gradually emerged like hibernating creatures, acknowledging the warmth.

"Rain's over," the woman said, taking her seat. "Outlook is much better."

"Thanks," I replied.

I returned to the hotel, remembering my thoughts of the previous evening, which had centred on finishing the hike. They were illogical; why walk all this way to quit with two days to go? I decided to carry on in the morning and fell asleep wishing for sunshine.

I checked the directions to Kirk Yetholm. East out of Peebles the route options were limited. The Cross Borders Drove

Road carried on south, in the wrong direction. A lack of trails made the decision to follow the B7062 to Innerleithen easy. After, I could pick up quiet country roads before skirting to the south of Galashiels, where I planned to pick up St Cuthbert's Way. I had two more nights and my crossing of Scotland was complete. Further plans depended on whether I returned home or carried onto the Pennine Way into England.

My tent arrived, and I left the hotel, returning to the café for breakfast before hoisting my pack and leaving town. A path by the roadside made for safe walking, and traffic was light. The River Tweed slipped past gracefully to one side and kept me company. It was in no hurry, and neither was I.

A short day of 12 miles beckoned to Innerleithen, which I'd pinpointed because it had a campsite with showers, and because the hotel in Peebles had sung the praises of a curry house called Saffron. I hadn't eaten curry so far in Scotland, and visions of a chicken jalfrezi with a well-done garlic naan kept appearing every time I closed my eyes.

I passed Kailzie and Cardrona, turned left at a junction onto the B709, and crossed a bridge over the Tweed. Innerleithen buzzed with cars and people returning home after work, and I made a mental note of the Saffron's location as I walked through the High Street. The Tweedside Caravan Park, as the name suggested, bordered the river. It proved difficult to find and required multiple requests for directions from the locals before I located it.

"Tent pitches are at the far end near the river," said the lady on reception, otherwise known as Sheila according to the brass name badge on her lapel. "The shower block is right in the middle, it's on your way, you can't miss it. Showers are free, love."

I smiled at families peering from the windows of their static caravans. They looked bored, and I, seemingly, offered the only entertainment that day. I pitched my tent on manicured grass, sheltered against a hedge in case the wind increased, then showered and left for town.

The Saffron was so tiny that, whenever a group of four customers entered to collect takeaways, the serving staff couldn't move. I counted seven tables – including mine, tucked in the corner sporting one chair for us single people with no mates.

"I've been dreaming about a chicken jalfrezi for days," I said to the waitress, smiling.

"We can turn your dream into reality," she replied, off the cuff.

"A well-done garlic naan too please – burnt bits all the better – no rice but a vegetable samosa starter would be great. Oh, and a glass of the Chenin."

"We have a half-litre carafe for just a little more, sir."

"Go on then."

I settled to read a discarded local newspaper and, once more, aware of my weakness but too pathetic to argue, made inroads into the carafe. My jalfrezi was excellent, and I paid, ignoring a craving to visit the pub on the way back to the campsite.

I lay on my stomach in my sleeping bag and selected the map app on my phone to research routes once more. My thoughts turned to Kirk Yetholm, but more importantly to what to do afterwards.

I had no idea how I'd feel when I arrived. Perhaps I'd return south? That looked likely. My tolerance levels were low, and all I yearned for was home where I could stay dry. This was familiar thinking: when hiking I often missed home comforts, and conversely, when at home, I dreamt of the trail.

Entertaining ideas, I followed the Pennine Way on the map. It's a 267-mile monster trail that has reduced many a hiker to their knees, and it left Kirk Yetholm heading south. I knew I could link up rights of way back to West Sussex should I wish (and have the strength to hike them). Also, numerous well-known hiking routes dotted the country.

A short gap between the end of the Pennine Way and the start of the Staffordshire Way could be linked together. A few miles in to the Staffordshire Way and I could switch over to the Heart of England Way, which finished right where the Thames Path started. The Thames intersected the Ridgeway, an absolute classic and rich in history stretching back thousands of years. It claims to be the oldest road in the UK. Avebury stone circle, near Stonehenge, lay at the western terminus – where more planning could link up with the South Downs, my local trail, where I wouldn't even need a map.

I found a disused railway bridge over the River Tweed by the campsite in the morning, saving me a long detour back through town. I flitted between the old railway line and the road until the Elibank and Traquair Forest presented a change of scenery. The road was still pleasant enough in itself, bumping along next to the river, with an occasional cottage tucked away in the trees.

I hoped for a coffee in either Ashiestiel or Peel, but both hamlets offered nothing. After Peel, a bridge crossed the Tweed but meant walking along the busy A707. I chose to carry on along the south side of the river, skirting the Yair Forest; then a mix of roads, broken up by dirt tracks, promised to deliver me to Abbotsford.

My mind raced, busy either plotting routes home or planning what I'd do if I decided to finish at Kirk Yetholm. To calm myself, I remembered my meditation and enjoyed two relaxing hours thinking of little, instead focusing on finding headspace, then utilising that space to find calm. I do this by counting to ten, then returning to one, before repeating. This encourages concentration and helps push other thoughts away. It's impossible to ignore these images, and they battle for attention, but being aware they are present helps to dismiss them. The result is long periods when my brain quietens and the interference ceases. When I stop counting, the mind remains calm, and so do I.

I found St Cuthbert's Way, which squeezed through the Eildon Hills. The Eildons comprise three peaks, the most prominent and northernmost being 1,385 feet high, with remains of a summit fort, occupied as far back as the Bronze Age. At the bottom, a huge Roman fort called Trimontium, named after the three hills, has now crumbled away to nothing. Eildon Mid Hill has a monument to the Scottish novelist and poet Sir Walter Scott.

St Cuthbert himself was a 7th-century saint who began his dedication to God at Melrose Abbey, where the route starts. It finishes 63 miles later at the Holy Island of Lindisfarne, where he was laid to rest.

I had already covered a good distance when I passed through St Boswells. The area is historically rich, dominated by the Romans, who built a long, straight route called Dere Street south of the village. It ran from the Antonine Wall all the way to York, around 146 miles further into England.

I'd clocked up 30 miles, and fatigue nagged at my legs. I was also ravenous. Dere Street cut a line through open countryside with no hope of a private place to camp, but I saw a forest east of Ancrum in the distance. After arriving at a good pace, I tucked myself away amongst the trees, spread my mat on the ground, and lay down, exhausted.

Shaws Butchers in Innerleithen had supplied me with some fresh beef, which they'd finely cut to reduce the cooking time. I drizzled a little oil in my pot and browned the meat, which spat and jumped about, threw in a chopped onion, then water and a stock cube. Towards the end I added a handful of rice to thicken the meal, salt and pepper and – of course – a shake of Tabasco. It was delicious, and I named it Dere Street Casserole.

Reflecting on my adventures, certain recipes dot the memories. On occasion, instead of the usual pre-mixed rice and pasta dinners I eat, I buy fresh ingredients and play around at camp. It doesn't happen often, because I'm so hungry I need food quickly, but I have fond recollections of the recipes I've concocted. I name them after the location and events surrounding that meal.

Meiringen Stew is a favourite, named after a town in Switzerland where I stayed in a park on the outskirts. I had a sachet of oxtail soup, a carrot, onion, and some sliced salami. Boiling the vegetables and adding the soup mix towards the end produced one of the best camp meals ever, and I still cook it at home sometimes.

On El Camino, I'd passed a derelict house by the side of the trail. The garden, unattended and overgrown, was bursting with herbs, and I helped myself to rosemary, sage, and thyme (hikers behind me that afternoon commented that I smelt wonderful). Further on, a farmer had finished lifting potatoes from his field and had missed a few stray ones, which I also commandeered. To my surprise and delight, an old woman just a little further on was feeding fresh sardines to her cat when I stopped to chat in broken Spanish.

"You like?" she said.

"Yes," I replied, "I love sardines."

With that, she filleted two of a generous size, put them in a plastic bag and gave them to me. When she also handed me a small plastic bottle of red wine, I couldn't believe my luck. I camped outside Castrojeriz, a beautiful town dominated by an impressive castle in the Burgos province of Spain. I made a fire and wrapped the potatoes, fish, and herbs in foil (which I often carried for such occasions) then left it to cook on the embers. After 30 minutes of wonderful aromas, I ate the lot. It's known as the Castrojeriz Herb Parcel.

Other memorable recipes didn't even need cooking. On the Pacific Crest Trail, day one into a nine-day stretch in the High Sierra, I'd bought a few fresh vegetables and fruit items when I resupplied in the town of Lone Pine. Pulling up in the evening by a glorious viewpoint, after a tiring day, I was so hungry I couldn't be bothered to cook. I pulled my food bag from my pack and tipped out the contents. Tearing open a pack of peppered jerky, I also had an avocado (wrapped in spare socks to protect it), blueberries, fresh mint, and grated Parmesan cheese.

Filling my pot with a selection of each ingredient, I shook over a little salt and pepper and a drizzle of olive oil. It was one of the best meals on the PCT, and I never managed to recreate it because I couldn't source all the ingredients. I call it the Lone Pine Dine.

Digesting my Dere Street Casserole, I lay back on pine needles, clasped my hands behind my head, and looked skywards. It was July 22nd, four weeks since I'd started, and I'd covered 550 miles. An intense evening light drew long shadows. Scotland still remained bright until gone 10pm. Treetops swayed back and forth, framing a dimming sky. I wondered if the Romans had camped in the same spot; perhaps they, too, had cooked something to eat and then relaxed, looking skyward.

Too tired to pitch my tent, and with the promise of a dry night, I spread out a groundsheet and my mat, then slid inside my sleeping bag. I fell asleep but woke several times, as if keen to check the stars. When I stirred the following morning, an ominous red sky flashed.

"OK sir, that's a single dormitory bed, for one night, today, July twenty-third at Kirk Yetholm. Your payment's approved. Have a lovely stay."

I thanked her and ended the call. Kirk Yetholm was reachable, only 13 miles distant. My perseverance had paid off – and it amazed me. I was astonished that, through all the physical demands I'd asked of my body, it had delivered. But, most of all, my mind was stronger than I thought.

The psychological battle was wearing me down. Skulking on one flank was my depression and the torments it threw. It represented a company of archers, drawing their bowstrings, lifting their aim, pausing, then releasing an arrow volley. My attempts to negotiate a peaceful withdrawal were met scornfully, so I defended myself, donning armour and raising a shield over my head.

I had battled, and still was, but my constant efforts on the hike – and indeed for the years prior – were reaching a conclusion. Tired, sick of the arrows, and worn out from being on the defensive, I could only protect myself for so long.

I left the forest and paused, checking the map for directions. Looking in the vague direction of Kirk Yetholm, my eyes settled on the horizon and a feature I'd missed before: a line of hills, tall and unmissable against a clear blue sky. The Cheviots. I sensed emotions rising. Sadness the finish was in sight but also confusion. I wanted to finish, to end the journey, and be free from the mental torment. Another part of me didn't, because under the surface of that depression lay happiness and the enjoyment of being outdoors for long periods. It was there, but inaccessible under a mask of sadness. That's the part that needed to continue.

My route was easy – St Cuthbert's Way went straight to Kirk Yetholm. After the village, and sticking to the plan of crossing Scotland, I had three miles to England. I decided to get to the border, then either return home or continue on the Pennine Way, depending on how I felt.

St Cuthbert certainly knew how to pick a route, I'll give him that. His choice to follow Dere Street made sense; it was direct. I wondered what the road's condition would have been like when he had travelled along it in the 6th century, 600 years after construction. I expect most of the paths he took to the Holy Island are still in use today. It feels peculiar treading on the same ground that countless others have also trodden on through time. I was walking in the footsteps of saints, Romans, and further back to the first travellers.

Reaching the River Teviot, I pondered how St Cuthbert had managed to cross. The current was weak, but the Teviot would have been impossible to ford – it was too deep and wide. I imagined a boat, perhaps charging a fee (or free to a man of God?), that would have taken him across.

Following the south bank of the Teviot downstream, I reached the confluence with the River Jed and kept alongside before climbing a flight of steps to the busy A698. Thankful for the straight road where I could see what was approaching, I dashed across and picked up a narrow road on the other side. St Cuthbert's Way shared the path with the Borders Abbeys Way, a 67-mile circular route connecting four ruined abbeys at Melrose, Dryburgh, Jedburgh, and Kelso.

I climbed, and the landscape rippled, warming up for the Cheviots, which loomed closer every time I glanced in their direction. I turned to view the Eildon Hills once more, their dominance of the surrounding area obvious. It was clear to see why they'd been chosen for military outposts.

Out in the open, I hiked through fields with the occasional forest. The sun shone relentlessly, and my body dripped sweat, attempting to cool. The hamlet of Morebattle offered a solution with a lovely pub, the Templehall Hotel. A pint of Coke with a generous helping of ice solved a raging thirst. Being just an hour from Kirk Yetholm, though, I decided to leave lunch until I got there.

The map was looking busy. I felt for the cartographer, plotting a mad series of contour lines, as well as fitting in the Roman legacy with fort remains everywhere. I'll bet a smile or two appeared at the hill names: Grubbit Law, Wideopen Hill (very apt for the highest point on St Cuthbert's Way), and Cookedshaws Hill. My favourite name was reserved not for a peak but for a burn draining off Grubbit Law. Considering the series of contours surrounding it, and its winding course to reach Kale Water, I'm not surprised they named it Weary Stream!

I wasn't smiling for long before I realised I had to climb up and over them all. After crossing Kale Water on a footbridge, a steep climb of 750 feet soon had me gasping before I topped out on the first, Grubbit Law. The going eased as St Cuthbert's Way bumped along up high until Cookedshaws, where a straight descent brought me back down to a minor road. Bowmont Water greeted me, which I followed to end up in between two villages: Town Yetholm looming over to my left, Kirk Yetholm and its hostel right over the bridge. It was a couple of hours before the hostel was due to open, so I walked up to Town Yetholm to grab a late lunch.

I could see the appeal and why people praise the two Yetholms. Town Yetholm centred on its main street, lined both sides with stone houses. Flowers coloured the front gardens, and I strolled past the War Memorial, turning up a path between grassy banks to the local shop. A local had informed me that a café called The View was plying its trade.

Sitting by the window, I sipped a coffee and picked at my panini. I watched people outside visiting the shop below, others resting on benches by the memorial or returning to their stone cottages. I pondered my next move, deciding whether to continue on the Pennine Way tomorrow or go home.

The planning was frustrating and challenging. First, I needed to trust my decision, but faith wasn't forthcoming. Once decided, I then had to rely on my erratic psychological state to carry on. Both rested on unstable foundations with the potential to weaken overnight. Much hinged on the weather, which was forecast to be a mixed bag for the next few days.

I left through the quiet village to find the hostel and turned right down the hill towards Kirk Yetholm. Bowmont Water was low as it followed a course through a meadow. A strip of flattened grass confirmed the footpath on my map, leading straight to the hostel, but I continued along the road to see Kirk Yetholm. The Border Hotel and pub was the only sign of life, with a few late-afternoon drinkers sitting outside squinting into the sun. It is thought that the legendary fell-walker Alfred Wainwright left money behind the bar to buy a pint for anyone finishing the Pennine Way. I wondered if the tradition was the same at the southern end.

The Pennine Way starts on the High Street opposite the pub and shares the route with St Cuthbert's Way for about a mile before they split. I stopped at the junction and looked up the road, which rose to gentle, green, rounded knolls, belying what was to come. As if the weather were convincing me to carry on, it was a textbook summer's day: leisurely white clouds floating across a mid-blue sky.

England was four miles away, and my head was spinning, trying to decide whether to continue on the Pennine Way. I had to reach the border so chose to continue in the morning and finish my hike across Scotland. Then I'd make a further decision.

I walked down the hill to the Youth Hostel, a pretty stone building sitting on the edge of the village. A note pinned to a notice board inside suggested that I find a bed and said the wardens would return at 6pm to check on me. Commandeering a corner bed in the downstairs dormitory, I realised no-one else was about. I made use of the complimentary teabags and sat outside with a cuppa, sucking on my e-cigarette.

"I don't know if those are wise or not?"

Startled by a woman's voice, I turned to see her gesturing towards my e-cigarette.

"Bloody things," I replied, smiling. "I agree, not ideal but better than inhaling burning chemicals. Are you the warden? Hi, I'm Fozzie."

"Barbara," she replied, shaking my hand, "yes I am, and this is my husband Frank."

Frank peered around her, pushing glasses up his nose to focus on me.

"Pleasure," Frank said. "Have you settled in OK?"

"Yes, and looking forward to a quiet night, and making use of your fantastic kitchen."

"Let me know if you need anything," he added.

I resisted urges to go to the pub and return to the store to buy wine. Instead, I checked the hostel fridge for leftovers (always a cheap way of finding food), which Barbara had thoughtfully placed all on one shelf, and labelled 'free to use'. There was an onion, a section of cauliflower, a carrot and – to my delight – a little leftover coconut milk.

"What are you thinking?" Frank asked, seeing me scratch my head and pondering where my curry powder had gone.

"Vegetable curry," I replied forlornly. "But I've lost my curry powder."

He said nothing, but raised one finger to his lips, and beckoned me to follow him.

"This is my secret stash," he said. "I can't leave it in the kitchen because it's good stuff. Most is organic, and some I even order from India. Everything you need should be in here."

He handed me a plastic container full to the brim with every herb a cook might need. Turmeric, cumin, chilli, cardamom, coriander – the contents went on.

"This is fantastic!" I cried. "You're a lifesaver and clearly a man who loves a good curry."

"I do, help yourself," he offered, "but keep it quiet. I don't want everyone knowing I got this stuff."

I considered it a great evening for ingredients sourcing, and free at that. I also found a tin of lentils lurking on a forgotten shelf.

"Expecting a sample!" Frank called from reception, which was only fair in the circumstances. An hour later we tucked in to my concoction, which was met with approving noises from Frank. I decided to call it 'the best things in life are free Yetholm curry'. Lengthy, but suitable.

I left for the Pennine Way in the morning, and the hostel wasn't even out of sight when I heard Frank's cries.

"Fozzie! Wait!"

I turned to see him attempting a slow jog towards me, raising his hand as if offering something.

I walked back to lessen his exertions.

"Forgot to give you this and saw you passing the window," he said. "Here."

I instinctively accepted the small plastic bag he handed me.

"I mixed you some spices. There's a little of everything in there, it's the basic mix I use at home. Tastes great!"

"Frank, thanks!" We shook hands and he walked back, still out of breath.

I stopped at the Pennine Way signpost and looked upward, grimacing. I sported a T-shirt under blue skies, but ahead lay an imposing cloud bank looking none too friendly. The Pennines are notoriously open, offering little shelter save a stone wall here and there, and a distinct lack of trees.

I checked that my waterproofs and umbrella were in my side pocket, enabling quick deployment, and strode up the High Street towards certain wetness.

This is a bad idea.

Halter Burn merged in from the left to join the path and act as a handrail. I bumped along the hills before turning to all 1,801 feet of Black Hag. The westerly wind increased, buffeting me, and as I turned a face full of rain forced me to close my eyes. I struggled up to the fence line south of Black

Hag, and, once I'd climbed over the stile, I knew my original adventure had finished.

The fence marked the border between England and Scotland. I placed one hand upon it, expecting someone to appear from nowhere offering congratulations and a signed certificate, but something dawned on me. I realised that I had come to Scotland to hike but also to get to grips with my unhappiness. I had hoped to finish with answers, but I hadn't. All I had were more questions.

I battled on through driving rain, the wind gusting and the noise deafening. I'd take a few steps forward only for the hand of the storm to thrust me back violently. There was only one thing for it: a forced retreat to the Bat Cave.

Grateful for the dry inside, I checked the phone while flexing my leg muscles to stay upright. I'd now come off Black Hag and had a steep climb up to the Schil, at 1,969 feet. I peered out, flinching at the waves of rain washing over the Schil's flanks.

Then, a saviour. A mile from the summit, two words on the map met my gaze, and I sighed appreciatively.

Mountain Hut.

Chapter 13

The Wettest Summer for Fifteen Years

As I lifted the latch, a gust tore the door from my hand, sending it crashing into the wall.

"Holy shit!" cried a guy inside. "You scared the fuck out of me!"

I couldn't stop laughing.

"Really sorry, mate," I offered apologetically. "The wind's insane!"

I checked myself and took stock; my poncho had held firm, my clothes damp just from sweat.

"I'm Fozzie," I said, wiping the dirt and water from my hand before offering to shake.

"Kieron," he replied. "Some storm! I only came from Kirk Yetholm today, and I ain't going any further. Damn, that's nasty out there." He huddled back in the corner.

"Yeah it is. Hiked from the village myself, and, if it's OK with you, I'm staying here too."

"Grab a pew and make yourself at home, making a cuppa, you want in?" he offered.

"Fantastic, thanks."

I was about to sit when the door swung open again.

"Shit!" we both shouted in unison, cowering, half expecting the shelter to collapse.

"Bloody hell!" cried the wettest-looking hiker I'd ever seen. He stood dripping in the doorway, arms outstretched in resignation, as water pooled beneath him.

"That's nuts! I thought I was going to die!"

Kieron and I said hello, and he introduced himself as Chris. We spent 30 minutes sniffing, wiping our noses on our sleeves, and putting on dry clothes and warm jackets. Chris fired up his petrol stove and held his hands over the flames, rubbing them appreciatively. Gradually, cups of tea heated our bodies, and as the hut's interior warmed, we began relaxing.

If you're familiar with the hills, then you'll know that 'mountain hut' can mean anything: from a tiny space constructed from frail-looking timber and a dirt floor with enough space for two people, to stone-built, five-room affairs with every modern convenience available. Ours was the former, although, to be fair, the construction offered precisely what was needed: shelter. The interior was rectangular, around 10 by 6 feet, with wooden benches on each longer side. Simple but perfect for a few hikers to sit out a storm. But it was getting late, and we weren't going anywhere. We'd figure out a sleeping plan – the hut was home for the night.

Basic it may have been, but we were grateful. It reminded me of the timber shelters on the Appalachian Trail, with three walls (the front was usually open), a pitched roof, and a raised sleeping platform. There are around 250 of them, spaced the entire length of the trail, and their simple design offers shelter from the elements.

People are surprised when I tell them about huts and bothies. They want to know why there's no running water, heating, or electricity. The answer is we don't need mod cons. Most hikers carry sleeping bags and warm clothing, and we'll either have cold food or a stove to heat meals and drinks. We source water where available – and electricity? It's not as if we have to plug in the TV or washing machine.

From one end of the scale to the other. On my mountaineering trip to the Swiss Alps, my stay at the Cabane de Moiry, an Alpine hut at 2,528 metres, was relative luxury. They even had light switches, as well as a dining room, full kitchen facilities, proper bedrooms, and a log fire. It shared similarities with huts in the White Mountains section of the Appalachian Trail, run and maintained by the Appalachian Mountain Club. Most of them charged a hefty amount, mainly because of the demand. However, a few offered a bed in exchange for work – washing the dishes and such tasks. It can be surreal during winter, with a blizzard ripping across the landscape outside, and you're inside sipping on cocoa and sitting by a roaring fire.

The Pacific Crest Trail has a handful of refuges over its 2,640 miles. Perhaps the most revered and majestic (in terms of scenery, not facilities) is the John Muir Hut. Again, it offers no amenities other than walls and a roof. What's incredible is the location. It's remote – and, at 11,955 feet, the views are unbelievable in all directions. I passed it at midday and needed to get to town to resupply, so my timing was off; otherwise I'd have slept there.

With a few hours to Kirk Yetholm, Chris had almost completed the Pennine Way, and he was in good spirits despite the weather raging outside. He filled us in on the rest of the trail, with stories of Roman ruins, quaint villages in which to grab a bite to eat, boggy sections, and UFO sightings.

"Any werewolves?" I enquired.

"Huh? Did you say werewolves?" Kieron said, looking through steamed-up glasses.

"Yes, apparently Cape Wrath is crawling with them. What's the deal with the UFOs?"

"All I kept hearing," Chris began. "I didn't notice anything unusual but everyone was talking about strange lights in the sky, mainly in the evenings. I shouldn't worry, you won't see them for clouds anyway." That was a fair point.

Kieron had finished eating a ham sandwich, with smoky bacon flavour crisps as a side. I was deliberating cooking a Mexican rice dinner with chicken pieces (which were a little smelly but only a day past their sell-by date). We both watched, curious, as Chris set up kitchen.

His stove was powered by compressed petrol. A friend of mine used to have the same model, and I never trusted it, mainly due to its alarming habit of flaring. I stood back, wiser from experience, as he pushed the plunger several times to prime it, then released the valve, allowing a dribble of petrol for priming. He lit this, letting it burn for a minute to warm the main fuel pipe, then opened the valve.

"Whoa!" he screamed, thrusting his head out of the way before his hat could ignite as well. He screamed and started smacking himself in the face, as if killing midges.

Kieron and I were in hysterics.

"You OK?" I cried, glancing at Kieron, who was beside himself.

"I think so," he replied, laughing. "I thought the flame caught my hair!"

As Chris glanced our way we noticed he'd lost an eyebrow. Losing one, as opposed to both, seemed unfair, and unfortunate.

"You'd pay ten quid for that in a beauty salon," Kieron commented, trying to control himself, then added, "Twenty quid for both of 'em."

We both observed, impressed, as Chris prepared his meal. Dehydrated food obviously didn't interest him, as he had brought various ingredients including fresh garlic, onions, chorizo, a can of tomatoes, and eggs.

"Shit, you've got a lot of food there," I remarked.

"Yeah, and I have to eat it. Even if I finish hiking tomorrow, my pack is too heavy."

After drizzling oil in the pan, he began by browning the onions and sausage, added seasoning, and stirred for 10 minutes before tipping in the tomatoes. He finished off by cracking two eggs into the stew. The hut smelt like a Spanish restaurant. Kieron had drool evident on his chin; we were starving, and glancing at each other in anticipation.

Before Chris had even muttered the words "There's heaps, you guys hungry?" we'd offered our bowls like Oliver Twist.

The night was uneventful. We slept poorly on the benches, scared of rolling off the narrow seats, which Kieron managed to do around 2.30am, much to my and Chris's amusement.

The sun blazed through the window in the morning, and I went outside to investigate as Chris, bless him, cooked us fried eggs. It was a glorious day, the Cheviots framed by a mid-blue sky from one horizon to another. Saturated ground, the remnants of the night's storm, squelched as I returned to the hut, packed, and said my farewells to the guys. They both wished me luck.

I made good progress, despite sodden feet. The Pennine Way was easy to follow, being one of just a few tracks crossing the Cheviots' expanse. I checked the map anyway and chuckled at the names: Butt Roads, Windy Gyle (very apt), Mozie Law, Brooming Crook, and Cottonshopeburnfoot. Dere Street cropped up once more near a hill called Brownhart Law, along with a series of raised banks, the remains of a Roman camp.

True to form, by early afternoon the weather deteriorated from poor to downright horrible. Thankfully, I'd reached the Redesdale Forest, which provided cover as I worked my way to the A68. The Byrness Hotel was shut, although I doubted whether I could have entered anyway as I checked my mud-splattered legs. Overnight accommodation, in whatever form available, was becoming a habit. Frankly, I was sick and tired of being wet. I'd had just three days of decent weather. Every time I managed to dry my gear out, it got soaked the next day. Nick's words of wisdom rang out:

You can push through many obstacles with determination, but the hardest is poor weather. It's belligerent, it will break you eventually.

Unable to remember the last time I'd worn warm socks or shoes, I was reminded there's nothing worse than putting on soaked ones first thing in the morning. My kit had been damp since day one. I'd managed to dry out at the hostel in Kirk Yetholm, but now everything was wet again. My recent habit of seeking shelter, and not pitching the tent, crept up once more, and I scanned the map for the next town. Bellingham was 11 miles away, and I discovered there was lodging there called Demesne Bunkhouse. I called to book a bed but had three or four further hours of hiking. It would be a late arrival.

My plan suffered before even reaching a decent pace. Just after Byrness, a quarter-mile section of track was sodden. Hikers appeared and warned me the going was wet, and, rounding a corner, I discovered the trail had vanished under a few inches of water. I splashed through as the rain became heavier and looked skyward for salvation, dreaming of Greek beaches and hot sand.

Redesdale Forest mitigated between me and storm clouds for six miles before I left the trees and squelched through the final section, finally arriving in Bellingham at 9.30pm. As I walked past two pensioners chatting under a shop awning, I caught wind of the conversation.

"Rain again," one said. "Every bloody day at the moment."

"Aye," the other confirmed. "Reg was saying it's the wettest summer for years."

I heard the other reply. "More of the same for the next week. Damn washout, fed up with it."

As I opened the door to the Demesne Bunkhouse and stood in the lobby dripping like, well, a hiker that's just walked for 12 hours in heavy rain, the owner, Chris, took pity.

"Blimey," was all he could muster, but that nailed it.

"Yeah," I replied. "Sorry for the acute wetness."

"Don't worry. Wettest summer for years apparently," he added.

"So I hear."

"There's a drying room next door. Please leave your boots in the lobby here, and feel free to use whatever bed you want upstairs. Kitchen's to the left, lounge is there," he pointed as confirmation.

I thanked him before climbing the stairs and commandeering my favourite place: bottom bunk, in the corner by a window. None of the four other beds were taken, and in the other dormitory only two were occupied.

I stood under the steaming hot shower for 20 minutes, and gradually a semblance of life returned to my body. Returning to my room, as I passed the lounge, a sorry-looking weather man on the TV gesticulated at the UK map behind him, which cowered under black cloud symbols.

"Records so far in the northern part of the country suggest it's the wettest summer for fifteen years," he said, resignedly, the third such remark I'd heard in as many hours.

I retreated to the Cheviot Hotel. The bar was busy, mainly with outdoors folk stocking up on calories and alcohol in preparation for the next day's assault in the hills.

"Sir, what can I get you?" the barmaid enquired.

"Pint of Guinness and a menu please," I replied, squeezing apologetically between two locals.

"You walking the Pennine Way?" one asked, sipping his beer.

"Yes."

"Bit wet up there isn't it, lad? They say it's the wettest summer for—"

I cut him off mid-sentence and smiled. "Fifteen years? So I keep hearing, and, judging by my hike, I ain't surprised. Cheers." I lifted my pint and toasted him before scanning the room for somewhere to sit. It was hectic, but one solitary guy, seeing me eyeing the other chair at his table, beckoned me over.

"Feel free," he said. "I had the same problem an hour ago."

"Thanks," I replied, relieved to sit down. I noticed him wearing Wellington boots, to which he smiled.

"Totally fed up with wet feet," he commented. "Been on the Pennine Way, soaked every day, so I bought these. Not ideal for hiking, but I ain't got no blisters and my feet are dry!" He chuckled, sipping his beer.

"How's the food?" I asked.

"Great. Anyway, I'm done in. Heading to the hotel to rest. Enjoy your meal, and your hike."

"Cheers."

I slept well, waking, not surprisingly, to the sound of rain on the skylights. I skulked beneath my umbrella through Bellingham. The Pennine Way went straight through this quaint place, which must afford good business for the locals. I found the Rocky Road Café, and, deciding I wasn't moving anywhere for the day because of the conditions, I sat, ordered bacon and eggs, and started reading the newspaper. I couldn't concentrate, picked at my breakfast, sipped on coffee, and pondered the situation.

Looking back on my hike afterwards, I realised that Bellingham was where I had made the decision to quit. I decided to carry on once the rain subsided, but, in all honesty, I wasn't fooling anyone; I'd had enough. My tenacity had deserted me. I had nothing left to give.

Despite completing two huge thru-hikes in the US and several in Europe, my mind had defeated me. I loved the outdoors, but after 31 days I was shattered. I decided to leave Bellingham and struck a deal. If I reached a point in the morning where I was still unhappy, and just trudging onwards purely from pride and stubbornness, then I'd quit.

"Fuck it," I said too loudly. A man looked up. "Sorry," I offered. I paid and walked out into the rain once more.

Returning to the bunkhouse, water streamed over gutters and poured from drainpipes, and waves washed over the flooded High Street. Hikers dotted the village, sporting colourful waterproofs, their heads tilted away from the downpour. They looked miserable.

I spotted a bike shop who, wisely, were advertising a well-known brand's range of waterproof gear. Impressed by the owner's marketing acumen, I checked the window display, noting an item of interest, and ventured inside.

"Hi," I said. "I've owned two pairs of those waterproof socks. They both leaked. Have they changed the design recently to convince me otherwise?"

"The older versions weren't great," he admitted, "but I've not had any returns on the current ones. I'd be surprised if you had any problems."

I paid and left. Back at the bunkhouse I removed the packaging and inspected my new gear, scrutinising the fabric and stitching as a horologist might inspect a wristwatch. I had doubts, remembering my first pair, which had failed on a mountain bike ride – strangely enough nearby, in Kielder Forest. I thought I'd learnt my lesson, but I bought another pair a few years later, which also leaked. I picked, prodded, pulled, stretched, and concluded there was only one test, and that was to wear them.

The morning never came, in terms of getting back on the Pennine Way at least. I ended up sheltering in Bellingham for three more days. The rain didn't let up, and every evening I sat in front of the TV expecting great things from the weather report. I berated myself for being so miserable; I should have been stronger. But I knew that if I ventured out, I'd be in trouble.

I spent most days huddled by the log burner in the lounge, a pathetic shadow of a man who not so long ago had thought nothing of pulling in 2,500-mile hikes. I'd convinced myself my hike back south was finished.

Finally, on day four of my stay in Bellingham, I woke to sunlight on my face. What a strange sensation. The forecast promised sun until the afternoon, when the best the weathergirl could offer was a 50 per cent chance of heavy rain, or a 50 per cent chance of no rain, depending on your outlook.

I packed hesitantly, glancing at the bus timetable I had picked up. Just in case, I knew the departure times to Newcastle and when the trains ran back home. Maybe the decision to quit had already been made.

My gear was dry for once. After time in the drying room, my clothes were warm, no longer heavy from moisture. After eyeing my socks warily, I pulled them on, lifted my pack, and left Demesne.

I kept pausing as I walked through the village. I felt pitiful, weak, and undecided. All I needed were 20-mile days and I'd be home in two months. My body revved, in first gear with the clutch slipping, the wheels spinning. But, my head, the engine's CPU, spluttered and misfired. One wouldn't work without the other. I wanted to scream in frustration.

Just let me do this! Leave me alone!

Doubts gnawing, I sat on a bench, trying to calm myself with shaking hands. I felt tears rising, and the corners of my eyes moistened; I fought them back and retreated to regroup in the Rocky Road.

"Thought you'd be long gone by now," the waitress said. "One last bacon and eggs, huh love?"

"That would be great, thanks."

I stayed in the café for two hours, making several false starts before reading the newspaper once more. Eventually, fed up with my indecisiveness, I left and followed the Pennine Way signs to the bridge over the River North Tyne. Traffic hissed on the damp road, and I cowered as trucks roared past, causing waves of water. After a mile I found the path that rose to Ealingham Rigg. As I jumped off the stile and landed, my feet sank in the soaked ground. I watched as my shoes disappeared under a murky brown mess, but my socks seemed to hold firm. Hesitantly, I splashed up to a trig point on the summit, using a communication mast as a visual bearing.

Sitting on the wall for a rest, I checked my feet, peeling off my saturated socks. As I put my hand inside each one, it was evident that, once more, their research and development department needed a budget increase. Dark clouds crept in from the west, suggesting the 50 per cent chance of rain was more like 100 per cent.

Even the map hinted that I should consider quitting. Unable to believe my eyes, I scanned over the place names: Shitlington Crags and Shitlington Hall, Sadbury Hill, Brownsleazes, and, best of all, High Moralee and Low Moralee! I stopped looking further for fear of seeing Giveup Hill, Quitandreturn Crag, or even a simple Turnaround Lane.

I wrung out my socks, gritted my teeth, and splashed off Ealingham Rigg downhill. By the time I'd reached a stile near Wark Common, as if by final verification, a storm cloud burst and a deluge of prime English summer rain smashed onto the Pennines.

I sat on the ground, took out my e-cigarette, and huddled under my umbrella. I don't know what the farmer who stopped his tractor must have thought. From his angle, all he could see was a pair of skinny, bare legs, topped with a silver poncho and umbrella, with clouds of 'Fruit Blast' smoke rising from beneath (I couldn't find any other flavours in Bellingham). He lowered his window.

"Everything OK, mate? You look like you've given up!" he called over the hum of the engine.

I smiled, raising my hand in resignation. "I have."

And with that, I turned right and traipsed back to Bellingham.

Chapter 14

A Year Later

It is no measure of health to be well adjusted to a profoundly sick society.
Jiddu Krishnamurti

Returning home after a hike is often difficult for me, but this time was different. I needed to get home. Travelling back, I reflected on my Scottish adventure while staring aimlessly out of the train window.

Physically I felt great – 550 miles of tough hiking is an efficient way of getting in shape. But my mind was a wreck, and psychologically I was exhausted.

Still unaware of the reasons for my moods, I carried on decorating (as I hated my job, that made the situation even worse). Despite weak attempts to improve my well-being, I tumbled downhill. Painfully aware of my decline, I didn't have the strength to fight.

I struggled with winter for the usual reasons: the lack of sunlight, drop in temperature, and deteriorating weather. I'd like to return as a bear. They've figured out how to escape the colder months and hibernate. One long sleep seemed appealing; I wanted to forget my woes for half a year and wake up in the spring.

Endeavouring to stay active, I walked daily and kept an eye on my nutrition. I meditated before bed, which proved beneficial – but, as always, I couldn't sustain the effort, nor, unfortunately, the results. The further I ventured into winter, the tougher it became. I went out less, made my excuses, and shied away from social engagements. At times it felt as if I *were* hibernating. I struggled through to January, then nosedived headlong into the pit once more.

Deeper than ever, I stopped looking up, the light just too far above, so tiny it barely resembled a speck. I shivered, cold, wet, and unable to see in the darkness. I felt the slimy stone and knew I couldn't climb out. Water dripped, echoing, the only company I had – but even that ceased, the walls swallowing all sound, and silence resumed.

As usual, I turned to drink. Seeking solace in alcohol had served me well before. It numbed the pain, relaxed the body, and I could handle the world for a few, short hours in the evening. Despite deciding against it every morning, I walked to the supermarket each afternoon to buy drink. I resisted the urge, determined not to make the trip again, but each day I relented, thinking I'd pop out for a bag of salad and avoid the Rioja shelf. Who was I kidding?

Each evening I finished the bottle earlier. To compensate I opened it later, but that didn't work, and I drank the lot. The last glass emptied sooner – 10pm, 9.50, 9.40. I bought more, thinking I wouldn't go over the limit, but before long, I'd uncorked another.

The alcohol fallout meant I couldn't sleep – or, at best, slept poorly. At 3am every night after frustrating attempts to drift off, I ventured downstairs and made a cup of tea, resisting the urge to glance at the drink cabinet. Sitting in silence, reading the paper, I returned upstairs and mustered a few, short hours of pathetic slumber.

I woke feeling as though I'd never been to bed. I stayed under the sheets as long as possible, waiting for enough light to fool myself into thinking it was May or June, then got up nursing a hangover and fumbled around in the drawer for the ibuprofen. This happened every morning for three months – or over 120 bottles of Rioja.

I'd discarded my vaporiser in favour of cigarettes. My lungs hurt, especially where I'd suffered with pneumonia. I tried to quit, but smoking and drinking were my escapes. I damaged my body, smoked too much, my alcohol habit neared an addiction, but I didn't care. Maybe a health scare was the wake-up call I needed. Perhaps death was an option? Shit, it would have solved a lot of problems.

When the wine stopped hitting the spot, I returned to smoking marijuana, which completed my return to and reliance on stimulants. Since returning from the Pacific Crest Trail, my life had flitted between long periods of extreme well-being when I took care of myself and shorter bursts when I fell spectacularly off the healthy wagon. The return to drinking wine, smoking weed and cigarettes was complete. Finally, my nutrition, normally excellent, suffered, and my commitment to exercise faded.

The one area I always focused on, despite whatever happened in my life, was my diet. I paid attention to what I ate because the body performs more efficiently with better fuel. Most days my plate was covered in salad, with plenty of raw vegetables, nuts, seeds, and good fats such as olive or coconut oil. Just once a week, as a treat, I'd consume meat, but usually sourced my protein from healthier sources such as fish, legumes, and eggs.

In the constant search to make my life bearable, I started eating poorly. Processed food, refined carbohydrates, beef, sweet treats, oven meals and more filled the fridge. The healthy options diminished. Even my route around the supermarket changed, my usual mental map now a distant memory. Before, I'd head to the produce lane first for fruit, salad, and vegetables, then turn in to the chilled section for juices. I'd veer right into the grains section to pick up lentils and legumes. To finish, I'd grab coffee, green tea, oatcakes, and cans of mackerel. But that was the past.

My new regime consisted of a weak nod to the me of old as I picked one bag of spinach, then filled my trolley with frozen curries, bread, chilled desserts, chips, and pizzas. Then I contemplated my choices in the drinks section – now wine boxes, not bottles. Oh, and don't forget tobacco on the way out.

I'd moved back in with my parents after the Pacific Crest Trail, a decision based on my need to save money to thru-hike more and concentrate on my writing. I remember every time I returned from a trip to the supermarket, slinking in quietly and unloading my junk food booty into the fridge before Mum noticed, for fear she might see. She'd give me strange looks when my plate – which used to resemble an allotment plot – degraded into the beige, yellow, and reds of a pizza, as I cupped my glass in case the red contents looked too obvious. I started buying screw-top bottles so she wouldn't hear me uncorking another.

My exercise regime disintegrated. Physical activity of any kind proved too much effort. Hell, I didn't even want to leave the house, let alone expend any more energy than necessary. I forced myself out on Sundays to do the usual nine miles, but there it stopped. No more mid-week walks, and my yoga focus every day soon became every two days, three, four – finally nothing. My bicycles seized up, collecting dust, and rust spots appeared on their chains.

Endeavouring to spend one part of my afternoon being physically active was normal, because I felt great for it. I'd try to walk for an hour, often two, or a run, cycle, or my usual yoga routine. Eventually, I did nothing. Your body knows when you treat it well; if the brain senses exercise, it rewards us with an endorphin rush. This feels fantastic, the body's way of forming a connection, a realisation that working out is beneficial, so we have the desire to do it again. But with no stimulation and little for my endorphins to look forward to, they too became depressed. No more whizzing through my head in a frenzy; instead they sat around watching TV, eating pizza, and drinking red wine too.

Truth is I was conscious of my actions and how they affected me. Despite my diet, lack of motivation, and the overall decline, my saving grace, as strange as it might sound, was being aware of my deterioration. I knew *exactly* what I was doing. I just didn't give a shit.

I continued chasing an unhealthy lifestyle that fuelled my mood. A vicious cycle – eat poorly, drink wine, and don't exercise. Indulge to escape, feed guilt because of indulging. Net result? Deeper despair, and yet I made no effort to change. I convinced myself that I loved being miserable.

I still didn't know I actually had depression. Something seemed wrong, but I consoled myself that in the spring I could begin to get on top of it.

Withdrawing from the world, but denying to anyone I'd left, I quit going out, refused to see anyone, and dreaded socialising. Invitations arrived but were declined: *Sorry, too busy, don't have any money, I'm too tired.* My friends aren't stupid. They had suspicions.

When asked what was wrong, I said nothing. A few persisted, insisting I'd been too quiet, that they hadn't seen me for ages. Something must be going on.

I'm fine.

The mood swings became erratic. If not let be, if I had no solitude, and people kept pressurising, I'd snap. I said many nasty things; I didn't care who I offended. I wanted them to stay away, to leave me alone – and if they didn't, I launched a warning shot. If it missed, the hint not taken, I loaded a bigger shell, improved my aim, and fired that.

When a friend is clearly unwell, most of us do our best to offer support. We offer help, call them more than usual, or send a text. With depression, or mine at least, I wished most of my friends would go away. I wanted to be alone, to revel in my misery. I had no desire to socialise or sit around a dinner table making small talk about who was pregnant, the colour of the new car, or voting on wallpaper choices for the lounge.

Nobody realised I needed that solitude. I screamed for it. Most of them, like good friends, refused to go away. This made the situation worse.

Just leave me the fuck alone.

I repeated my pleas that I was fine to no avail. I became offensive so they'd stop pestering – encouraged arguments and resorted to unpleasant comments. Gradually, they backed off.

However, I'm grateful that most of them stuck with me.

Elina kept in touch occasionally, but I still dismissed the notion of depression. My downhill slide continued, falling to a precipice, getting closer to a sheer drop every day. A cliff edge at which there'd be two choices: either plunge or stop myself. Dropping over meant the end, whatever that entailed, but a handhold offered the last hope. I hadn't prepared for the fall, but sometimes I yearned for it; I needed confirmation my life was out of control. When I reached that point, I promised to grab that section of rock and hold on, climb back, and begin rehabilitation. I'd take it right to the finishing line.

The precipice arrived in March. Once again, I struggled out of bed with a hangover and contemplated another day fearing the pit. I sat at my desk, looking out the window, watching rain trickle down the pane. I stared at overcast skies until my eyes glazed.

Sometime later I awoke but with an unnerving sensation that I hadn't actually slept. I'd passed out; how long for didn't concern me, but the shaking did. Looking at my hands, in disbelief and bewilderment, they shook uncontrollably. The lack of control scared me, and I began crying. Not a weak, pathetic attempt but a full-on blow out. The house was empty, so I released it all and sobbed. I buried my face in trembling palms, felt the trickle of tears flow down my wrists and dribble onto my forearms.

That Monday in March was the worst day of my entire life. I let go. I started falling towards that cliff edge. I saw the drop, thousands of feet below, and my stomach clenched.

A freezing wind ripped across me. I gazed at the world below, a cliff face tumbling to a huge, icy lake. The waters reached the horizon and melted into angry skies, black clouds raced over the landscape, lightning bolts blazed.

I grabbed the rock and held on, and there I dangled.

Fall or climb, Fozzie. Fall or climb.

The phone rang. I didn't want to talk to anyone, but as I glanced at the screen and saw her name as my finger hovered over the accept icon, I knew she was the only person I could speak to.

"Hey," I managed.

"Hi. What are you doing? You don't sound good. You OK?"

"No."

"What's going on?"

"I'm in a very bad place. Something's wrong and I don't know what to do."

Four hours later I sat in the doctor's office. Initially, they said they were fully booked; then I told them it was urgent, that I was worried about my mental health.

"Come in at 5.45," they said.

Depression is like chest pain. You know it's serious when the doctor wants to see you the same day.

I have a great doctor. I like him because medication is the last choice he considers; he looks at other possibilities first. Holding myself together, I relayed my experiences.

After asking me a series of questions, presumably to reach a diagnosis, he listened intently as I told him of my despair, feelings, and sadness.

"It sounds like you're suffering from depression," he concluded.

"Really?"

"Yes. Everything you've told me is indicative of mental illness."

"So, what now?" I replied, expecting him to write a prescription for antidepressants.

"We have three prongs of attack," he said, leaning back in his chair. "The most important first, you should see a counsellor. Then we look at your lifestyle, and finally, if we're not making headway, we consider medication, but I'd do that reluctantly. How much are you drinking?"

The question surprised me.

"Sorry? How do you know I drink?"

"Ninety per cent of my patients who suffer with depression consume a lot of alcohol. There's a strong link."

"I do a bottle of wine most nights. Sometimes more," I replied sheepishly.

"Unless I think you're a danger to yourself, this discussion doesn't leave the room, Keith. What about stimulants?"

"I smoke weed every night."

"Anything else?"

"No, cigarettes, but that's it."

"OK. I'm going to refer you to the local NHS therapy team, they'll be in touch. For now, and I know this will be difficult, but I want you to stop drinking and smoking marijuana for three months."

"What?" I returned his gaze, shocked. "Do you realise how hard that will be?"

"Yes, I do. But believe me, far easier than coping with depression."

I looked away, ashamed.

"Drink is connected. We're not sure about weed, but we suspect there are links as well."

"I'll try."

I shook his hand and turned to leave.

"Oh, and Keith, one more thing. Ask for help. More people than you think are in the same boat."

After the diagnosis, I held on to the fact I had a mental illness during the dark days. Believe me, it's a strange position to be in – conscious of suffering from depression, where the only consolation is realisation. But finally comprehending the truth ended up being my saviour. When I fell in the pit, instead of wondering why, *knowing* I had depression helped – despite how far away reality seemed.

And I had a lot of rough days. A friend told me he suffered, and likened the experience to a rock being thrown into a pool many years before. In the months afterwards, the waves threatened to overwhelm him, and 20 years later he still feels the ripples. I agree, although every few weeks, as the water calmed, someone threw another rock.

I didn't stop drinking or smoking, but I made a deliberate effort to cut down. *Cutting down* is the old excuse addicts offer instead of stopping. I returned to vaping, stopped doing weed in the week, but I couldn't shake the wine habit.

Two years after Scotland and I'd gone from deep depression to being able to hold it together, just. Treading a line between improving my lifestyle and returning to habits of old, I still indulged in the addictions, but I had managed to quit cigarettes.

I wanted antidepressants during the hard times. Luckily my doctor disliked them and convinced me otherwise. My issue with depression medication is this: to treat a condition, we need to find the cause. For example, with a stomach infection, we isolate the bacterial strain and prescribe the appropriate antibiotic. If we cut ourselves, the hospital staff inspect the wound and, depending how deep, will either dress it or apply stitches.

Now with depression, and indeed most mental illnesses, we've yet to find the cause. The experts have ideas – they'll pinpoint events such as relationship break-ups, or a death in the family, but they still haven't discovered the root of the problem.

We're medicating against an unknown. If we're unaware of what's causing the problem, how can we treat it? I realise many people lead better lives because of antidepressants, but they may never escape them. I worry about the repercussions in years to come from prescribing drugs for a condition we don't understand.

In the summer, I decided that time away would help. I booked two holidays, one to Crete in May, the other to Majorca in September. On both occasions, I woke up on the day of departure, looked at what I needed to pack, couldn't face it, and called the airline to cancel. In the space of a few years, I'd gone from 2,500-mile hikes to being incapable of packing a suitcase.

Then my mum fell ill. After being diagnosed with cancer in October, she underwent chemotherapy and then radiotherapy. Eight months after, in June, her condition deteriorated rapidly. My sister, dad, and I nursed her at home until her admission to a hospice, where she passed away a week later.

I dreaded the funeral. We were exhausted after looking after her, making arrangements, and dealing with the official paperwork. Losing her was painful enough, but then I had to spend the day with other people I hadn't seen for months, including my best friends. Even a few people seemed daunting; now I had to socialise with 100. For days leading up to the service, I struggled to cope. How I found the strength to hold it together I don't know, but it turned into a wonderful remembrance.

Some of her ashes are scattered on the South Downs, the rest in the countryside near Parwich, Derbyshire, where she spent time growing up, and a place she loved all her life. We spread wild flower seeds. As I write this, in the depths of winter, I long for spring when they bloom, and I can sit with her and enjoy them.

We knew the prognosis wasn't good in her final days. My smoking and drinking were under control – just – because I had to hold it together for her, and my mental state remained reasonable. I dreaded the day she'd leave us. I thought I'd be unable to handle it, but, in the end, her death helped me.

When I saw her battling to live for another day, I took strength from it. My mum fought hard for her life, and I was throwing mine away. She tried everything to survive; my attempts were pathetic.

She died in July. I stopped writing and spent more time outside, going for walks most days, and thinking. The summer blazed. I soaked up the sunlight, my body grateful for the exercise. Feeling stronger and confident, in September I again made arrangements to go to Greece with a friend. I planned to spend a week with her, then attempt a hike I'd been planning for years: three weeks crossing Crete.

I had two great weeks, doing simple things; sunbathing, walking, reading, talking, eating, and drinking. But when she left for the airport I broke down and cried. The thought of being alone, returning by myself, and some history between us knocked me back. I spent a couple of days making weak arrangements for my hike and decided I didn't have the strength to see it through. I returned home the following day.

Towards the end of September, I noticed various social media sites promoting 'Go Sober for October'. It suggested abstaining from alcohol for one month, either for personal reasons or to raise money for charity. My depression rampaged again. I was weak, convinced alcohol was to blame. Sad and miserable because of my failed hike in Crete, the winter months loomed, and alarm bells began ringing loudly. My condition could potentially turn nasty. I decided to do something about my life.

I set two goals for October: to stop drinking and smoking weed for the entire month. Despite overindulging often, I never found pot addictive and was confident I could quit. Breaking my alcohol habit would be different.

Changing my evening patterns to escape old habits, I left for the gym later so it interrupted the usual time for my first glass of wine. After returning home, I busied myself with preparing dinner and focused on the cooking, sipping fruit juice instead.

I stayed away from the friends whom I always used to drink and smoke with. They soon noticed. One of the repercussions of abstaining from stimulants is those who try to persuade you otherwise. You'll get little sympathy or support from your drinking buddies or your weed smoking group when quitting. They don't want you to because they'll be on their own. They realise it's wrong to be leading that lifestyle, but they seek validation from others so it excuses what they're doing. I stopped seeing them; when they asked why, I told them I needed time out and I'd see them in a few weeks.

I managed that first day, October 1st, without a drink. In fact, I got through the entire month.

When I hit November, I was flying. My body felt better, but my concentration and focus rocketed. Instead of hazy, weak efforts to get through my writing targets, I ripped through them with ease. I attacked my tasks vigorously, with a single-minded determination. Assignments I usually found difficult to understand suddenly became clear.

More importantly, for those four weeks I never suffered badly with depression. Whether the lack of alcohol or weed was responsible I didn't know, but it was a hell of a coincidence. I decided to stay with my plan. My drinking and smoking friends made several attempts to recruit me back into their ranks.

"But it's November," they cried. "You said you'd just quit for October!"

"I did say that, yes. And I also said if it went well, I'd continue. I'm not saying I'll never drink or smoke again, but I'm going to ride the wave and see what happens."

"Have a smoke on Friday, we'll buy some beers, get tuned up, and hit the PlayStation."

"No."

Two weeks later, I did intentionally partake in some wine, and weed. I hadn't relented, but I was curious to observe any repercussions. Call it research if you will. I felt good on Sunday, albeit hungover, but on Monday and Tuesday my depressive symptoms returned. Not severely, but enough to take notice. I continued to stay away – and so did my moods.

I wrote a blog, where I posted about my depression. Wary of judgement, the response, in fact, was amazing. I received messages from people I'd never communicated with offering support. A subscriber, Brad Devine, replied.

'Hi Fozzie,

'I just wanted to share with you a quote from one of my favourite authors and someone I think was an absolute visionary, Aldous Huxley (*The Doors of Perception*, *Brave New World*). It reinforces the truth that it is perfectly *normal* to feel the way we do. It's something to celebrate.

'The real hopeless victims of mental illness are to be found among those who appear to be most normal. Many of them are normal because they are so well adjusted to our mode of existence, because their human voice has been silenced so early in their lives, that they do not even struggle or suffer or develop symptoms as the neurotic does. They are normal not in what may be called the absolute sense of the word; they are normal only in relation to a profoundly abnormal society. Their perfect adjustment to that abnormal society is a measure of their mental sickness. These millions of abnormally normal people, living without fuss in a society to which, if they were fully human beings, they ought not to be adjusted.'

Brad continued:

'The Indian philosopher Jiddu Krishnamurti arguably put it better when he said:

'It is no measure of health to be well adjusted to a profoundly sick society.

'But personally, I like the way Huxley breaks it down, it's funnier. I hope this helps Keith. For me it has always been a source of strength, of pride even, because as much as it might be hard sometimes, I'm glad that I don't sit comfortably in this twisted society. I'd worry more if I did!'

Brad's message made me realise that mental illness, even in today's civilised society, is still stigmatised. How uncaring is our culture when we feel ashamed to admit something in fear of how others will respond? At that point, I realised it wasn't me that had a problem – our culture did. I began telling my friends about my depression, unashamed, and they responded sincerely. In fact, many replied and said they endured mental illness too.

I still suffered setbacks, but of less severity and frequency. If removing alcohol and weed had resulted in positive changes, what else could help? With a focused mind, I started looking at other areas where I could make improvements and break away further.

The most important thing I did was seeing a doctor and getting a diagnosis. I realised that, although I'd never escape depression, making changes in my life could control it.

I spent hours researching nutrition, exercise, meditation, supplements, and other areas. I noted what I thought relevant and believed could work for me, and I put them into action. I continue to tinker with that list; I add points as I go along, or advice from others. Now, months on from that first day of October, I still get bad days, but they're rare, and I deal with them. The steps I have taken mean I'm in a great place.

I don't spend much time in the pit any more. I prefer sitting in the woods listening to birds.

In the last chapter, I'll tell you what I discovered, how it affected me, and what you can do to break away from depression if you suffer with it.

Appendix

How I Finally Won – My 15 Methods to Treat Depression

L et me make one serious point before I start. It's
disclaimer time; I'm no expert in psychology, nor am
I a doctor, and I don't offer therapy.

This list is the result of research, trying, testing, and
tweaking. What helped me may not work for everyone, and
if you discover something beneficial that isn't listed, please
tell me so I can share it with others.

I mean that – contact me via my website:
keithfoskett.com.

I advise seeing a doctor before making any lifestyle
changes like starting an exercise regime, changing diet, or
adding natural supplements. Some natural supplements
shouldn't be taken if you have certain medical conditions, or
if you are taking other medication. For example, St John's
wort shouldn't be taken with many drugs, so check before
taking.

This advice is just that, *advice*. Please take responsibility for your own well-being, and be careful.

I accept no responsibility for the guidance or information given here.

The following points keep my depression under control. It may never go away, but I can cut both the frequency and severity of my symptoms by taking these steps.

There are 15 suggestions in order of priority, number 1 being most important, in my opinion. You could find number 7 works better than 5, but, as I said, it's what I found effective.

Adjust, tweak, add, and take away as you see fit.

The most dramatic improvement happened after I stopped drinking alcohol – but I have friends who have depression who barely drink. When I cut down using cannabis, I noticed the positive change, but again, some of my mates still indulge and tell me they don't have mental health issues. We're all different.

When trying these methods, the effects could be different for you. Some could work, others won't.

When researching depression, I noted the experiences of other individuals and what approach they used. I refined those results to the list below, which I consider the most important points, and my advice as well. There are countless other beliefs out there, but it's impossible to do everything. Do some research. Note the ideas that appeal and which you could commit to. I have the occasional drink, have days where I eat poorly, and occasionally I should get more sunlight – but I endeavour to do as many as possible on a daily basis.

If you're serious about addressing your condition, take whichever points you can, and try to include them in your lifestyle.

Do I still have depression? Yes – I don't believe it will go away. But, after making this list, and practising it, my symptoms are less frequent, and not as severe.

Remember the rock and pool? Well, that rock has been thrown, but it's down to you how you handle the ripples.

What I can say is this. If you do suffer from depression, and take these points on board, I'm sure you'll feel a better person for doing so.

Step 1 – Don't be Rude to the Cashier
AKA – Be Aware

Step 2 – It's Easier than we Think
AKA – See a Doctor

Step 3 – The Repercussions of Procrastination
AKA – Act Now

Step 4 – Crossing the Street
AKA – Therapy

Step 5 – Turn on the Bat Light!
AKA – Look on the Positive Side

Step 6 – The Feel-Good Myth
AKA – Stop Relying on Stimulants

Step 7 – Service the Engine
AKA – Nutrition Overhaul

Step 8 – Animals make Simple Connections
AKA – Exercise

Step 9 – Bag of Spinach Hard to Swallow?
AKA – Herbs, Minerals and Supplements

Step 10 – Bacteria *is* Good!
AKA – Improve Gut Health

Step 11 – Toxic Friendships
AKA – Remove the Negatives

Step 12 – Full Fat or Semi-Skimmed?
AKA – Meditation

Step 13 – Animals aren't Stupid
AKA – Seek Sunlight

Step 14 – Make the House Smell Great
AKA – Beneficial Oils

Step 15 – Final Points

Step 1 – Don't be Rude to the Cashier
AKA – Be Aware

I've been a moody sod the past few years. Completely oblivious to my mental state, I accepted the bad days as normal. I never looked for explanations – does anyone? Do we ever stop to think why we snapped at our colleague for no reason?

My downfall was the supermarket. Often, shamefully, and being stressed, I was more than curt to the cashiers on occasion.

Such events can seem normal. We don't think to ask why.

Depression may encourage irrational behaviour. If we're aware of the signals, they can be addressed.

Consider the following signs – and if they seem familiar, or tick a lot of boxes, see step 2.

Psychological symptoms

Psychological effects include feeling low or sad. You may have little hope for the immediate or long-term future. Poor self-esteem and crying are common, as are feelings of guilt. At work, concentrating can be difficult, and you may have a short attention span. Some cannot find any motivation for even simple tasks. Suicidal thoughts are common.

Physical symptoms include fluctuating weight and variations in appetite. Aches and pains, a lack of energy, and erratic sleep patterns are not unusual.

Finally, mental illness causes social anxiety as well. Problems such as underperforming at work, avoiding friends and social situations (especially group activities), neglecting hobbies and interests, and strained relationships affect sufferers.

Step 2 – It's Easier than we Think
AKA – See a Doctor

I've had many jobs. A few years ago, I started labouring on building sites, and the manager used to put me on the painting tasks because I had a talent for them. It didn't take long for me to figure out that I could earn better money and enjoy more freedom if I went self-employed.

I agonised over whether I should be my own boss for two years. Would it work? Could I source enough clients? How much would the equipment cost me?

Eventually, I committed and started my own decorating business. It turned out to be one of the best moves I ever made, and I flew through the first year. I was super busy, increased my earnings, didn't have to answer to anyone, and had more control over my life.

My only regret was not making that decision earlier. After I did, everything became obvious, especially the fact that I shouldn't have delayed. I should have committed from the start. My fears amounted to little, my doubts unfounded.

Major decisions are difficult, and we do over-analyse them. Seeing a doctor if we suspect mental illness is tough because admitting it is hard in itself.

Sometimes it was easier living in ignorance. Did I really want to know if I had a mental illness?

Please, if you think you have depression, make the decision and call a doctor.

Depression *is* dangerous, and common. Many of us suffer with it without realising, and it can be debilitating.

Practitioners recognise the symptoms and will investigate further. If they suspect mental health issues, they'll offer advice, and have methods at their disposal to help.

Picking up the phone is a brave move. The appointment could be emotional, and you may cry. It can be tough opening up.

It will be worth it.

Step 3 – The Repercussions of Procrastination
AKA – Act Now

I've just downloaded an album called *Combat Rock* by the Clash, a punk band from way back. The 80s was a musically radical and diverse decade. I hated punk at the time and couldn't understand why many were raving about it, but now I love it.

It first caught my eye ten years ago. I had albums by the Stranglers, the Damned, and Green Day. I looked at the Clash and secretly knew *Combat Rock* would be great, but I kept putting it off.

Eventually I bought it. It was fantastic; some albums take several listens to gauge, but I loved it from the first play.

Here's the moral of the story. I spent many years without that band. I could have been listening to them since I got into punk in the 90s, but I procrastinated. If I'd parted with my cash a long time ago, my Clash enlightenment period could have started earlier.

As hard as it is to reach decisions when we're depressed – and indeed to act on them – try not to procrastinate.

Act now. The more deliberation, the longer we spend in the pit.

Step 4 – Crossing the Street
AKA – Therapy

Experts in depression tell us the most important choice is to see a therapist.

For me, it was last on the list.

I have to admit they haven't totally worked for me. I've seen a few counsellors over the years for various reasons, and a couple of hypnotherapists as well, and I don't rate those experiences.

However, one lesson I learnt helped me enormously. Cognitive behavioural therapy (CBT) is a method of managing a condition, including depression, by changing the way we perceive certain situations.

My therapist used this example. Imagine walking on a busy street. We see a friend approaching and intend to stop and say hello, but they cross the road. Now, we can react in two ways. First, we think they crossed because they wanted to visit a shop. Or, possibly that side was sunnier?

Or, we imagine they did notice us, but were avoiding contact. Then, we compound the situation and begin wondering why they didn't want to see us.

Perhaps they had issues with you refusing the party invite they sent. Maybe they didn't agree with what you said to their brother in the pub last week. Before long you've dissected the event too much and concluded they don't like you. Not only that, you spend several days rolling it around in your head and making it worse.

Just because someone crossed the street!

The depressive mind isn't interested in the positives. It bombards us with negatives. It wants to bring us down, and, to do that, it creates destructive emotions.

It's like being clamped in medieval stocks while depression throws rotten apples, and there's no escape.

Don't let my experience with counselling deter you. I did learn one important lesson – CBT. It made a massive difference, and that, alone, warranted seeing a therapist.

Therapy takes time; weeks, usually months. For that reason, it makes sense to start early.

Step 5 – Turn on the Bat Light!
AKA – Look on the Positive Side

The Mayor of Gotham City has a major problem. The Joker is holding several high-ranking officials hostage, and they're roped over a huge vat of bubbling, corrosive lime-green acid. In 24 hours, the ropes will be cut.

What does he do? He turns on the Bat Light, of course! He asks for help!

The moral is: talk to people, ask for support.

Simple? Yes.

Easy? No.

The last thing I wanted to do was talk to anyone. I couldn't even *see* anyone.

My thoughts were so irrational that I refused to believe someone else could be in my position, let alone millions of others. No-one could help me.

But, they did.

A lot of people are suffering the same hardships, many have come out the other side, and I found a few of them. I felt great opening up. It helped me because they knew what I was going through. They shared the steps they took to fight depression; what worked, what failed, ideas I was unaware of, approaches I hadn't contemplated, and methods I never thought relevant.

I concealed my condition for over a year. Not because I felt ashamed, I just didn't need to broadcast it. I dealt with it on my own.

However, what surprised me, when I did start opening up, was people's reactions. They held my gaze, listened intently with little emotion, then simply told me they understood completely, and that they, too, had been battling depression for years.

I was dumbstruck.

And they helped me, making valuable suggestions. Better advice than the therapist even!

Try finding those people and talking to them.

Turn on the Bat Light!

Step 6 – The Feel-Good Myth
AKA – Stop Relying on Stimulants

From reading my story, you know the importance of this step.

Stimulants provided me with relief from depressive feelings, but it was short-lived, and the fallout merely compounded the problem.

I drank a lot of alcohol, smoked too much weed and tobacco, and ate unhealthily. I felt great in the evening, when I usually indulged. My pain numbed, my comprehension of the world changed to somewhere I could relax. Then I had to face the morning hangover, the coughing, stomach aches, and, obviously, a deeper depression.

You're already aware of the repercussions. I have the occasional drink still, but I'm proud to say that I'm now drinking just two bottles of wine a month instead of 40. Once every couple of weeks at most I indulge in a little weed. Smoking tobacco, or using an electronic cigarette, are long gone.

Quitting or drastically cutting down on stimulants was the single most effective step I took.

Step 7 – Service the Engine
AKA – Nutrition Overhaul

This point is so significant I nearly entered it at number one. It's not as important as realisation, seeing a doctor, or getting therapy, but I could put forward a compelling argument to move it to the top three. Fact is, you'll be in a better place to tackle this step after completing those first six.

This relates to the body, and centres, as the title suggests, on what we eat.

For years, food has fascinated me. My diet consists mainly of vegetables, fruit, whole grains, legumes, nuts, seeds, fish, good oils, and a host of herbs and spices. I occasionally eat what I consider 'bad' foods like meat, a doughnut, or a pizza. I take supplements and drink three pints of water a day.

We cannot possibly get our minds in reasonable working order if our bodies are neglected.

Let's use a car analogy. Two people each have the same vehicle. One fills it with top-grade fuel, changes the oil every six months, and services it at 10,000 miles, using the best parts. It gets cleaned most weeks, and the levels are checked, along with tyre pressures. That person drives sensibly and doesn't thrash the engine.

The other driver uses the cheapest fuel they can find, rarely checks the fluids, and tops it up with reconditioned oil. The tyres get air if they look flat, the services take place at twice the recommended intervals, using poor parts, and they drive their car hard.

Now, you know the question I'm going to pose as well as the answer.

Which vehicle runs the most smoothly, efficiently, and with fewer breakdowns?

Exactly.

It's the same with our bodies. We're far more complicated and wonderful than an engine, but we are just that. If we neglect ourselves, particularly by consuming poor fuel, we won't run optimally.

If our body isn't firing on all cylinders, neither will the mind.

It's difficult to focus on what we eat – especially when depressed – because comfort food makes us feel better. But we need a whole host of nutrients to perform at our best. Our bodies can't function properly by subsisting on a diet high in unhealthy fats, carbohydrates, or poor forms of animal-derived protein, with no fruit or vegetables.

This is fact – science has proved it.

To get on top of mental illness, we have a simple choice. Continue eating inferior food and drinking soda, I won't judge you, we're all free to make our own decisions.

Or, consider a diet overhaul.

Many of us still doubt that changing what we eat will have any difference. Here's another story that illustrates my point. In medieval times, one of the biggest killers wasn't cancer, or obesity, but scurvy. In fact, between 1500 and 1800, it's estimated this disease killed at least two million sailors. Mariners were particularly prone, especially during long periods at sea.

The solution was simple: consume more vitamin C. But the main source of it is fresh fruit and vegetables, which have a poor shelf life, and those foods perished quickly on board.

If eating oranges cured one of the biggest diseases ever, consider what proper nutrition can do for us.

I enjoy a burger and fries as much as anyone, I still drink wine occasionally, and I'll be a sad man when they stop making Snickers. But I'm strict with how often I partake.

The odd bar of chocolate, beer, or ice cream won't cause depression, but it's worth reducing poor food choices. I feel terrific as a consequence of my diet, and my mind is in great shape as well.

Evidence points towards specific foods helping in the fight against mental illness. The classic example is the Mediterranean diet, consumed in such countries as Italy, Greece, and Spain, where fewer people suffer from depression. A diet rich in vegetables, fruits, fish, whole grain, and olive oil reduces the risk.

Nuts, in particular, are worth mentioning. Low levels of selenium are linked to a higher risk of depression. Brazil nuts, as one example, contain a lot of this mineral.

Scientific studies have shown that a lack of vitamin B can affect our temperament. Look for foods containing folic acid, particularly green, leafy vegetables, preferably only lightly cooked.

Research has suggested that omega 3, found in fish oils, may improve mood. Eating food like tuna, mackerel, or salmon, flax or rapeseed oil, or munching on pumpkin seeds will help.

Serotonin is a chemical in our brains that makes us feel good. Low levels of this substance are connected with depression. It is derived from tryptophan, which our bodies can't produce, so we need to source foods that do contain it such as cashews, tuna, salmon, beans, and seeds.

Finally, drink enough water. Arguably this doesn't come under nutrition and demands a point to itself, but I treat it as something I consume, so it fits.

I try to drink three pints a day. After spending a lot of time peeing, it takes a while to get used to. My trick to avoid constantly going to the toilet is having an entire pint three times a day as opposed to sipping gradually – one when I wake up, then around lunch, and finally early evening. My body requires this amount ideally; you might need the same, maybe less, maybe more. The internet has loads of information on how much to drink according to our individual needs.

Why? We cannot function without it. It is essential for life and for the body to work efficiently. I wouldn't go as far as saying it treats depression, but there's little point in eating healthily, quitting stimulants, exercising, and everything else if you're dehydrated.

Think of food as your fuel and water as the engine oil.

Step 8 – Animals make Simple Connections
AKA – Exercise

Keeping fit is good for us.

In my early thirties, I realised that I couldn't have a strong mind without a healthy body, and that when I exercised I was happier.

Ever seen a dog in the park just running around, changing direction quickly, jumping and generally having the time of its life? That's because it is!

Animals make simple connections – they eat when hungry, and if they're tired, they sleep. And they realise that exercise is great.

When we work out, the brain releases endorphins, sometimes known as the feel-good chemical. I see it as our minds' way of providing a connection, telling us we should get active more often.

I try to exercise every day. It could be a walk, a run, swim, or a cycle. Lifting weights and yoga are favourites too. I'm not a full-time fitness weirdo – some days I don't want to do anything at all.

But I know that, when I do make the effort, I'm in a better place afterwards.

Just like that dog in the park.

Step 9 – Bag of Spinach hard to Swallow?
AKA – Herbs, Minerals, and Supplements

There's evidence to suggest that adding supplements to our diet can improve mental health. People lacking in iron, for example, often sense weakness and fatigue, amongst other signs. If someone is deficient in iron, taking an iron tablet will bring them back to a healthy level. I speak from personal experience on this one.

A lack of zinc in the diet makes us more susceptible to infection, can thin hair, and can even impair hearing. Adding a zinc supplement can ease these symptoms.

Certain minerals relieve depression. The best known is St John's wort, used for centuries to treat many conditions, including mental illness.

You've probably heard of omega 3 fatty acids – they've had a lot of press attention in recent years. Adding them to our diets is thought to improve brain function.

B vitamins, in particular B12 and B6, are important for brain health. They help produce chemicals that influence our moods. Low levels are known to cause depression.

Vitamin D, known as the sunshine vitamin, may help reduce symptoms. As with vitamin B, not getting enough vitamin D is linked with mental illness. Exposing our bodies to sunlight is a great way of getting this nutrient, as well as eating foods such as eggs, sardines, and cod liver oil.

Referring back to point 7 – nutrition overhaul – remember we can add foods to our diet rich in these minerals and nutrients, although taking a tablet is more convenient.

Believe me, an iron pill is far easier than a bag of spinach.

There are confusing signals with herbs, vitamins, and supplements. Many sources claim they help; others dismiss them as folklore. A few are banned in the US but freely available in Europe and vice versa. The above are examples of the popular remedies, but there are lots more. Do your own research – what works for me may not for you.

Step 10 – Bacteria *is* Good!
AKA – Improve Gut Health

Excuse the pun but I've decided to look after my stomach bacteria. Don't know why, just a gut feeling!

This point was a late addition to my regime, and I only became aware of the advantages recently. It's receiving a lot of attention, and rightly so.

The gastrointestinal tract is home to a host of microbes collectively known as the microbiome, and the highest concentration is in the gut. Broadly, there are two types of bacteria present: good and bad.

Scientific research claims we're not getting enough of the good variety, and too much of the bad, which can cause problems from weight gain to a weakened immune system, poor skin condition, and – wait for it – a decline in mental health.

Inadequate nutrition, such as too many highly processed foods, alcohol, caffeine, sugar, and artificial sweeteners, weaken the microbiome.

Eating healthier food and incorporating more sources of good bacteria can help. There are two types of these foods: prebiotic and probiotic. Probiotics contain the good bacteria themselves, and prebiotics are foods that nourish them.

Probiotic yoghurts have been in the supermarket chiller section for years. Others include fermented foods such as sauerkraut, kimchi, kombucha (I hadn't heard of most of them either), some cheeses, and apple cider vinegar.

It's important when considering fermented foods to *not* buy the pasteurised kind. All pasteurisation does is kill off the good bacteria, which is the food industry's way of getting rid of anything they think could harm you. Ironic isn't it? These foods are great for your gut health, but because they contain that terrible word *bacteria*, the food industry must err on the side of caution, and the law, and pasteurise them.

Some people need to be careful when consuming them, for example those who are pregnant.

Prebiotics – which I liken to fuel for the probiotic – feed the good bacteria and keep them healthy. Prebiotic is dietary fibre. Major sources are chicory root, dandelion greens, Jerusalem artichokes, garlic, onions, leeks, asparagus, bananas, and oats. Eat raw if possible; cooking can reduce the benefits but is inescapable for foods such as Jerusalem artichokes.

To sum up, introduce probiotics to top up with good bacteria, then feed them prebiotics.

Step 11 – Toxic Friendships
AKA – Remove the Negatives

I read one of those 'Earn a Million Dollars in a Year' books once. I learnt very little from it, and neither did I make any money, but I did take away an important point.

To be successful, sometimes we have to remove harmful aspects from our life. Negatives drag us down, acting as obstacles. The positive stuff is on the other side; we just need to break the barrier to reach it.

I needed to be brutal. I was in a bad way and battling to overcome it. Drastic action was called for.

One of those measures was ending friendships, two of them to be exact. I realise now they were toxic; neither party gained anything, and my life is better without them. It took a lot to exclude those people, but I realised it was the right thing within a week once I'd done it.

An example. One friend was a heavy drinker, smoker (tobacco and marijuana), and ate enough fast food to keep every burger bar in the village in business. When we hung out, that's all she wanted to do. When I changed my habits, she questioned it and became frustrated and angry towards me.

Do you remember me talking about my friends who criticised my decision to stop smoking and drinking and did their best to discourage me? Same principle; another friend needed people around him, including me, to indulge in those practices because it validated his failings. He didn't give a damn about me as long as I ate pizza, drank vodka, and got high.

I told him how I felt. He refused to accept it, and clearly didn't care about me, so we parted company.

Sometimes we have to remove the negatives.

Step 12 – Full Fat or Semi-Skimmed?
AKA – Meditation

Coincidence can be persistent, can't it? I always find it amazing that I go through life oblivious to something, then it crops up three times in one week.

For example, has a friend ever recommended watching a movie you've never heard of? Two days later the same film appears in a newspaper article. Then, it actually comes on the TV and turns out to be fantastic. You wonder how it passed you by all this time.

That was me and meditation. Up until my mid-thirties it evaded me. No magazine articles, friends hadn't mentioned it, and it never cropped up on the TV. Somehow, I completely missed it.

Even when I did stumble across it, I dismissed it as mumbo-jumbo – and I have a very open mind. What a big mistake.

Ask ten people who practise meditation why they do it, and we'd get ten different answers. The common advantages include stress reduction, increased feelings of happiness and well-being, and improved concentration. It aids restful sleep and benefits cardiovascular and immune health.

For me, the main benefit is that it reduces my head noise.

What's head noise? Ever tried to sleep, and your mind won't shut up, spinning pointless stuff around that doesn't even need contemplating?

All you want is eight hours' kip, while it's demanding to know how much milk to buy in the morning. Semi-skimmed or full fat? One pint or two? From the supermarket or the little shop?

Irrelevant isn't it? But our minds will over-analyse until we eventually pound the pillow several times, scream "Argh!" and get up to make a cup of tea (or drink a pint of water!).

It takes practice, and commitment, but the noise can be stopped. My approach is to breathe in and count one, then breathe out. Inhale again, count two, exhale and repeat to ten, then begin again.

This exercise focuses the mind away from distractions. Ideas still enter my head and demand attention, but I simply acknowledge them, let them go, and return to counting.

Semi-skimmed or full fat? I don't care, three, four, five...

What's for dinner tonight? Doesn't matter, six, seven...

My favourite time is when I'm outdoors exercising and want to soak up the countryside without distractions. After a few weeks' practise, I found I could walk for an hour and meditate simultaneously.

It's a head workout. I've explored physical exercise and how it benefits the body – well, it's the same for the mind. Give it a workout, it'll love you for it.

And, wait for it, meditation is closely connected to improving mental health!

Step 13 – Animals aren't Stupid
AKA – Seek Sunlight

Ever owned a dog or cat? If so, you'll know they love sunlight. My cat, for example, spends most of the time outside. When it's cold, he seeks sunshine, and it's amusing to see how he does it.

When I let him out in the morning during winter, the first thing he does is sit on a raised part of the garden. That place gets the sun first as it rises, and there he stays, facing directly into the light, eyes closed, soaking it up. If he could sigh with satisfaction he probably would.

When he's inside, he knows what time daylight enters the lounge and where it shines, so that's where he lies.

Animals rely more on awareness than us humans, because they act on instinct and don't question as we do.

Getting enough sunlight has many benefits, of which one

is alleviating the symptoms of depression. Come the summer months, I try to get two hours of exposure every day, usually by walking. It boosts levels of serotonin, the body's happiness hormone.

Please take suitable measures in the sun: apply sunscreen, avoid the hottest times around midday and early afternoon, and restrict the time spent exposed to it.

Step 14 – Make the House Smell Great
AKA – Beneficial Oils

I've always used oils in my house. Up until recently I was unaware that certain types have different effects on us. I like them because they smell great, although they can also have a calming effect. People regularly comment that my place smells wonderful.

Turns out oils can improve our mood.

My favourite is frankincense. Other types linked to the treatment of depression are ylang ylang, lavender, roman chamomile, and bergamot.

Use them by adding a few drops to water, and heat the liquid with an oil burner, or try a diffuser.

Step 15 – Final Points

Remember, if the experts are right, we can't yet cure depression, only reduce the severity and frequency. It's a sobering thought that those suffering from a mental illness might never fully recover.

However, please, don't give up. We're not trying to cure it; the goal is to live a relatively normal life, with fewer symptoms, occurring less often, and less severely.

Using the above methods, I have managed to improve my condition to a level where, often, for days, I forget I even suffer. At worst, I'll have the occasional bad day. From where I've come from during those dark times in Scotland, the improvement has been immense.

Remember there are other possible approaches as well; I haven't covered everything here.

I can sense those of you out there with depression muttering.

"This is all great. It's fine you saying go and join a swimming class, cook a vegan feast, cut out wine, or maybe buy a Clash album. But I can't even get out of bed."

I hear you. I *was* you.

One approach is taking small steps, incorporating the changes slowly. The first week try staying away from alcohol on Monday and Thursday. The following week abstain on Sunday as well, and so on.

The next day, call the doctor. After that, accept his suggestion of seeking therapy, and make an appointment for a couple of weeks' time. Gradually implement those healthy eating ideas, and start to take a daily walk after work.

Small steps are more achievable, one each day, or even every week.

We may never completely balance those highs and lows, but, steadily, we can reach a stage where our depression is manageable – and we can lead a relatively normal life.

Printed in Great Britain
by Amazon

21476282R00163